Nuclear Engineering Division 2019

AA009160

Held at the 2019 AIChE Annual Meeting

Orlando, Florida, USA
10-15 November 2019

ISBN: 978-1-7138-0536-6

Printed from e-media with permission by:

Curran Associates, Inc.
57 Morehouse Lane
Red Hook, NY 12571

Some format issues inherent in the e-media version may also appear in this print version.

Printed with permission by Curran Associates, Inc. (2020)

AIChE
120 Wall Street, FL 23
New York, NY 10005-4020

Phone: (800) 242-4363
Fax: (203) 775-5177

www.aiche.org

Additional copies of this publication are available from:

Curran Associates, Inc.
57 Morehouse Lane
Red Hook, NY 12571 USA
Phone: 845-758-0400
Fax: 845-758-2633
Email: curran@proceedings.com
Web: www.proceedings.com

TABLE OF CONTENTS

(242A) CESIUM REMOVAL FROM TANK WASTE SIMULANTS USING CRYSTALLINE SILICOTITANATE AT 12% AND 100% BED HEIGHTS 1
Amy Rovira, Sandra Fiskum, Jarrod Allred, Heather Colburn, Margaret Smoot, Andrew Carney, Reid Peterson

(242B) MESOFLUIDIC EXCLUSION OF LARGE PARTICLES FOR NUCLEAR WASTE APPLICATIONS 2
Leonard Pease, Carolyn A. Burns, Timothy Veldman, Jason Serkowski, Nathan Phillips, Richard C. Daniel, Michael J. Minette, Xiao Ying Yu, Matthew Fountain

(242C) PARTICLE BACK TRANSPORT MECHANISMS IN CROSSFLOW FILTRATION WITH NUCLEAR WASTE SIMULANTS 3
Philip P. Schonewill, Richard C. Daniel, Carolyn A. Burns, Rick W. Shimskey, Reid A. Peterson

(242D) A CONVECTIVE-DIFFUSIVE FORMULATION FOR CROSSFLOW FILTRATION OF NUCLEAR DEFENSE WASTES 4
Richard Daniel, Philip P. Schonewill, Reid A. Peterson

(242E) DESIGN AND TESTING OF AN IODINE AND TRITIUM CAPTURE SYSTEM FOR USED NUCLEAR FUEL TRITIUM PRETREATMENT PROCESSES 5
Robert Jubin, Barry Spencer, Stephanie Bruffey

(242F) AGING KINETICS OF AG-EXCHANGED MORDENITE AND AG-FUNCTIONALIZED SILICA AEROGEL IN THE PRESENCE OF OFF-GAS STREAMS 6
Seungrag Choi, Yue Nan, Alexander Wiechert, Austin Ladshaw, Sotira Yiacoumi, Costas Tsouris, Lawrence L. Tavlarides

(242G) TOWARDS INVERSE DESIGN OF NANOPOROUS MATERIALS FOR ISOTOPE SEPARATION 7
Musen Zhou, Jianzhong Wu

(242H) BLIND SOURCE SEPARATION IN RAMAN AND ATR-FTIR SPECTROSCOPY: A PROCESSING CASE STUDY 8
Giovanni Maria Maggioni, Stefani Kocevska, Ronald W. Rousseau, Martha A. Grover

(309A) EVALUATION OF CHEMICAL ADDITIVES FOR GLYCOLATE DESTRUCTION IN HIGH-LEVEL RADIOACTIVE WASTE CONDENSATE IN THE DEFENSE WASTE PROCESSING FACILITY 9
Dan P. Lambert

(309B) TANK CLOSURE CESIUM REMOVAL AT SAVANNAH RIVER SITE 11
Terri Fellinger, Aubrey I. Silker, Mark Keefer

(309C) LARGE CEMENTED GIBBSITE AGGLOMERATES IN ALKALINE NUCLEAR WASTE AT THE HANFORD SITE AND THE IMPACTS TO REMEDIATION 12
Jacob Reynolds, Jason Page, Gary Cooke, Beric Wells

(309D) PROCESSING OF ACTINIDE MATERIALS FOR REUSE/DISPOSAL 13
Sharon Robinson, Brad Patton

(309E) IN-SITU RAMAN AND ATR-FTIR SPECTROSCOPY FOR MONITORING OF MOLECULAR SPECIES IN LOW-ACTIVITY NUCLEAR WASTE 14
Stefani Kocevska, Giovanni Maria Maggioni, Martha A. Grover, Ronald W. Rousseau

(309F) ULTRA-MICROPOROUS MATERIALS FOR EFFICIENT REMOVAL OF KRYPTON FROM NUCLEAR REPROCESSING FACILITIES 15
Sameh Elsaidi, Mona Mohamed, David Hopkinson, Ahmed Helal, Ju Li, Praveen K. Thallapally

(309G) THE MIGRATION LAW OF NUCLIDES IN STEAM REFORMING REACTION OF SPENT RADIOACTIVE RESINS 16
Qi Song, Yao Yang, Jingyuan Sun, Zhengliang Huang, Jingdai Wang, Yongrong Yang

(385A) DESIGN AND OPERATION OF A SYSTEM FOR THE PREPARATION OF A MOLTEN CHLORIDE NUCLEAR FUEL SALT 17
Justin McAllister, Michael F. Simpson

(385B) NEUTRON RADIOGRAPHY FOR THE DETERMINATION OF MOLTEN CHLORIDE VISCOSITY, DENSITY, AND HOMOGENEITY 18
Matt Jackson, Marisa Monreal, Scott Parker, Alexander Long, Sven Vogel, Bjorn Winkler

(385C) ELECTROCHEMICAL CORROSION MEASUREMENTS IN MOLTEN FLINAK SALT 20
Suhee Choi, Andrew Strianese, Michael F. Simpson

(385D) THERMODYNAMIC ANALYSIS OF MINOR COMPONENTS IN MOLTEN SALT MIXTURES 21
Richard L. Fitzhugh, John Harb, Matthew Memmott

(385E) METAL DISSOLUTION IN MOLTEN SALTS FOR REDOX CONTROL 22
Olivia Dale, Suhee Choi, Michael F. Simpson
(489A) HABIT 2.1: ASSESSING CONSEQUENCES OF CHEMICAL RELEASES ON CONTROL ROOM HABITABILITY CONSIDERING THE IMPACT OF A DIVIDING STREAMLINE 23
Thomas O. Spicer, Casper Sun, Syed I. Haider
(489B) NATURAL CIRCULATION IN A VERY HIGH TEMPERATURE REACTOR (VHTR) 24
Dinesh V. Kalaga, Apoorva Rudra, Masahiro Kawaji
(489C) SPECIATION OF FUEL AND FISSION PRODUCTS IN MOLTEN SALT SOLUTIONS FROM AB INITIO MOLECULAR DYNAMICS 25
Austin Clark, Wan Luan Lee, Andrew Russell Solano, Tyler Williams, Gabriel Meyer, Granite J. Tait, Ben C. Battraw, Avenly H. Hansen, Stella D. Nickerson
(489D) DENSITY FUNCTIONAL THEORY STUDY OF THE TRITIUM DIFFUSION, DESORPTION, AND TRAPPING ON THE γ-LIALO$_2$ (100) SURFACE 26
Ting Jia, David J. Senor, Yuhua Duan
(489E) FUNDAMENTAL STUDIES OF TRITIUM DIFFUSIVITY IN CARBON DOPED DEFECTIVE γ-LIALO$_2$ PELLETS 28
Hari Paudel, Yuhua Duan, David J. Senor
(489F) MODELING OF PLUGGAGE AND SPATIAL DISTRIBUTION OF PARTICULATE MATTER DEPOSITED IN THE SAVANNAH RIVER SITE H-CANYON PROCESS VESSEL VENT FILTERS 30
James E. Laurinat, Michael G. Bronikowski
(489G) SOLVENT EXTRACTION FLOWSHEET-LEVEL MODELING AND SIMULATION WITH RADIOLYSIS 46
Valmor F. De Almeida, Kevin Lyon, Taha Azzaoui
(551A) MODEL OF HELIUM BUBBLE DENSITY AND GROWTH IN TUNGSTEN FROM EQUATIONS OF STATE 47
Karl D. Hammond, Dimitrios Maroudas, Brian D. Wirth
(551B) ON THE ONSET OF 'FUZZ' FORMATION IN PLASMA-FACING MATERIALS. I. SURFACE MORPHOLOGICAL EVOLUTION MODEL 48
Dwaipayan Dasgupta, Robert Kolasinski, Dimitrios Maroudas, Brian D. Wirth
(551C) HELIUM AGGREGATION AND SURFACE MORPHOLOGY NEAR SURFACES AND GRAIN BOUNDARIES IN PLASMA-FACING TUNGSTEN 49
Karl D. Hammond, Zhuocen Yang, Dimitrios Maroudas, Brian D. Wirth
(551D) ON THE ONSET OF 'FUZZ' FORMATION IN PLASMA-FACING MATERIALS. II. EFFECTS OF SUBSURFACE BUBBLE DYNAMICS, TEMPERATURE, AND ELASTIC MODULI OF DAMAGED TUNGSTEN 50
Dwaipayan Dasgupta, Robert Kolasinski, Sophie Blondel, Asanka Weerasinghe, Dimitrios Maroudas, Brian D. Wirth
(551E) MECHANISMS OF RUBIDIUM AND KRYPTON RELEASE FROM ZEOLITE GETTERS 51
Amir M. Mofrad, Karl D. Hammond
(551F) DEFECT INTERACTION ENERGETICS AND STRESS STATE OF PLASMA-EXPOSED TUNGSTEN 52
Asanka Weerasinghe, Lin Hu, Dwaipayan Dasgupta, Karl D. Hammond, Brian D. Wirth, Dimitrios Maroudas
(551G) THERMOMECHANICAL PROPERTIES OF PLASMA-EXPOSED TUNGSTEN 53
Asanka Weerasinghe, Brian D. Wirth, Dimitrios Maroudas
Author Index

579475 Cesium Removal from Tank Waste Simulants Using Crystalline Silicotitanate at 12% and 100% Bed Heights

Augmented pretreatment is planned near the Hanford waste tanks in skid-mounted platforms termed Tank Side Cesium Removal (TSCR). A promising Cs ion exchange material, crystalline silicotitanate (CST) in an engineered bead form is under investigation to support the TSCR pretreatment flow sheet for Cs removal from the tank waste supernate. Pacific Northwest National Laboratory has been tasked with demonstrating the unit operations for the Low Activity Waste Pretreatment System with actual Hanford tank waste. Testing was requested by Washington River Protection Solutions to better define IX processing using Low-Activity Waste Pretreatment System (LAWPS) prototypic IX unit operation process steps at full height and medium height. Column testing at both the medium (12% of the full bed height) and full height scales was conducted to evaluate process variables and scale up performance of Cs exchange onto the CST. Nominal process conditions used 5.6 M Na simulant processed at 1.8 BV/h at 20 °C. Process variables included 1) increased process temperature to 35 °C, 2) increased Na concentration to 6.0 M, 3) added organics to the 5.6 M Na simulant, and 4) flowrate changes. The kinetics of Cs exchange were examined for each of the variable simulant test matrices. These studies were intended to provide results and experience using the CST in a full height column format and assess impacts of multiple process variabilities.

571948 Mesofluidic Exclusion of Large Particles for Nuclear Waste Applications

A key element of nuclear waste processing and treatment unit operations is the separation of suspended particles by size. For example, tank side cesium removal (TSCR) systems used to treat nuclear defense waste stored at the Hanford Site are working toward a requirement to remove particles larger than 10 μm, and the Hanford Waste Treatment and Immobilization Plant (WTP) pretreatment facility requires removal of particles larger than 310 microns. In these two example cases, removing large particles out of waste streams is important to protect the operation of downstream cesium ion exchange columns and the facility design basis for mixing and waste transfers, respectively. This presentation discusses a new mesofluidic separator that leverages fluid dynamic streamlines in column arrays to exclude large particles from waste streams without moving parts. Mesofluidic systems leverage microfluidic principles but operate at industrially meaningful flow rates (e.g., gallons per minute). The particle size range separated is tunable and tailored by design. Separator prototypes have been developed using metal and plastic 3D printing. With these advanced manufacturing prototypes, separation has been demonstrated in turbulent multicomponent slurry flows up to 90 gpm, thus far, for multiple separator orientations. In sticky nuclear waste simulants, the mesofluidic separator loses $<6\%$ of flow in 7 hours and is projected to lose $<13\%$ of flow over three years without back flushing or cleaning of the device. In contrast, the dead end filters currently anticipated for TSCR particle separations lose $>60\%$ of flow in 7 hours. The mesofluidic separator would operate at low pressures for many DOE waste challenges. This new separation technology has potential for broad applications across the DOE complex, in commercial nuclear reactors, and in cleanup of environmental waste waters.

570282 Particle Back Transport Mechanisms in Crossflow Filtration with Nuclear Waste Simulants

Crossflow filtration is a planned unit operation in the Hanford Waste and Immobilization Treatment Plant Pretreatment Facility and has also been considered for use in other pretreatment facilities on the Hanford site. Crossflow filtration may be used across a significant range of solids concentration, from a few hundred parts per million to 20 wt% undissolved solids. The wide range of possible concentrations, combined with the chemical complexity of the Hanford waste, introduces challenges in performing *a priori* estimation of filter fouling and production rates, *i.e.*, filter flux.

More specifically, a persistent disconnect has been observed between crossflow filtration models in the literature and Hanford waste or waste simulant filtration data; this complicates the prediction of long time performance using macroscale physical and chemical parameters. Most models forecast that a cake quickly forms on the filter surface and reaches a steady-state thickness within minutes or a few hours. In general, Hanford filtration does not follow that trend; either a steady-state is not quickly achieved or the models over-estimate the steady-state flux (which is often only a pseudo steady-state) by factors of two to three. To probe this difference, historical Hanford waste and waste simulant crossflow filtration data were examined against a series of proposed back transport mechanisms. Across the examined data sets, shear-induced diffusion was found to best describe crossflow filter performance.

576703 A Convective-Diffusive Formulation for Crossflow Filtration of Nuclear Defense Wastes

Crossflow filtration (CFF) is a solid-liquid separation technique that will serve as a key enabling technology in the disposition of nuclear defense wastes stored at the Hanford Site in the U. S. state of Washington. Studies evaluating technical sufficiency of CFF for separation of nuclear defense wastes observe a persistent loss of filter performance that continues throughout >100-hour batch processing times and that is not recoverable through filter back-flushing. Loss of performance, when coupled with the time-intensive and operationally complex chemical cleaning operations needed to restore filter performance, increase the time and cost associated with waste disposition (and could be rate limiting). Attempts to model fouling of CFF elements by Hanford waste solids using crossflow correlations available in the literature and semi-empirical have met with limited success. Literature models predict a rapid achievement of filtration steady-state not observed in Hanford prototypic CFF; and semi-empirical approaches cannot be extended to different operating conditions and provide no insight into the underlying mechanisms effecting performance loss. These models focus on convection driven fouling of >10 μm particles. However, Hanford wastes exhibit gel polarization when filtering at high-solids loadings, suggesting that diffusion or a diffusion-like mechanism controls foulant mass transport to and from the filter. The current study develops a convective-diffusive framework for evaluating filter fouling dynamics of Hanford waste systems. Application of this convective-diffusive framework to filter fouling dynamics measured for a dilute (~5-wt% solids) high-level waste slurry suggests that relaxation from a gel-polarized state dominates the earlier filter performance loss while the longer-time losses appear to be controlled by a diffusive-like mass transport of solids within the filter membrane.

572052 Design and Testing of an Iodine and Tritium Capture System for Used Nuclear Fuel Tritium Pretreatment Processes

Abstract will be supplied after approval by ORNL. - Completed 4/18/19.

Design and Testing of an Iodine and Tritium Capture System for Used Nuclear Fuel Tritium Pretreatment Processes*

R. T. Jubin, B. B. Spencer, and S. H. Bruffey

Oak Ridge National Laboratory, P.O. Box 2008, Oak Ridge, TN 37831, jubinrt@ornl.gov

During the reprocessing of used nuclear fuel, volatile radionuclides will be released into the off-gas of a processing plant, including ^3H, ^{14}C, ^{85}Kr, and ^{129}I. The management of tritium becomes complex once it enters the aqueous portion of the facility (i.e. the dissolver and downstream processes). One potential way to simplify the management of off-gas streams could be to incorporate an efficient tritium pretreatment (TPT) step in which the UO_2 fuel is oxidized either by air or by NO_2 before dissolution. Oxidation of the UO_2 fuel matrix results in the release of tritium into the off-gas stream. Upfront removal of tritium from the fuel in a pretreatment step, followed by the abatement of tritium, can minimize or eliminate the distribution of tritium throughout the plant, also decreasing or eliminating the need for tritium capture on multiple off-gas streams. The oxidation of the fuel can be accomplished using air, oxygen, or other oxidants such as NO_2. The use of NO_2 in an advanced TPT (ATPT) system allows the oxidation to be performed at lower temperatures and may also result in the quantitative release of iodine from the fuel.

The ATPT off-gas will contain up to 75 vol% NO_2 gas. Silver nitrate–impregnated alumina (AgA) has been identified as the most promising iodine sorbent, and type 3A molecular sieves (3A MS) or silica gel have been identified as the most promising tritium sorbents to treat ATPT off-gas streams [1]. An initial series of tests were conducted in fiscal years 2017–2018 [2] with the intent of demonstrating the recovery of iodine and tritium using these sorbents from a recirculating gas stream that simulated the NO_2 recycle loop envisioned for the ATPT system. These initial tests showed low tritium recoveries and significant variations in the iodine recovery on the AgA. Ultimately, following disassembly of the system, it was concluded that corrosion within the system was at least partially the cause of the poor recovery for both species of interest. One of the conclusions drawn from those initial tests pointed to the reconsideration of the materials of construction.

Based on that recommendation, a new system has been designed and built that uses high nickel alloys wherever possible. The system will be tested in mid-FY 2019 under air and NO_2 TPT conditions. The results of those tests will be reported in this presentation.

1. Spencer, B. B., S. H. Bruffey, J. A. Jordan, and R. T. Jubin, *Design of a Tritium and Iodine Removal System for Use with Advanced Tritium Pretreatment*, Report nos. ORNL/SPR-2017/116 and NTRD-MRWFD-2017-000311 (Oak Ridge, TN: Oak Ridge National Laboratory, 2017).
2. Jubin, R. T., J. A. Jordan, and, S. H. Bruffey, *Testing of an Iodine and Tritium Removal System for Advanced Tritium Pretreatment Off-Gas*, Report nos. ORNL/SPR-2018/15 and NTRD-MRWFD-2018-000199 (Oak Ridge, TN: Oak Ridge National Laboratory, 2018).

577722 Aging Kinetics of Ag-Exchanged Mordenite and Ag-Functionalized Silica Aerogel in the Presence of Off-Gas Streams

Iodine-129 is one of the volatile radioactive species in the off-gas streams generated from the spend nuclear reprocessing system, and it is essential to remove ^{129}I vapor because ^{129}I has a long half-life (1.6 x 10^7 years) and can be fatal to the human body if released to the environment. Radioactive iodine vapor can be effectively captured by silver containing adsorbents such as silver exchanged mordenite (Ag^0Z) and silver functionalized silica aerogel (Ag^0-aerogel) used in this study. However, the aging impact of off-gas streams on the adsorbents is one of major issues in the removal of radioactive iodine vapor during off-gas treatment. The iodine loading capacity on the silver containing adsorbents will decrease over time due to the chemical and physical changes to the adsorbents when exposed to off-gas streams. In this study, the aging impacts of off-gas streams on the kinetics of iodine capture on both Ag^0Z and Ag^0-aerogel were investigated. The adsorbents were aged in dry air, humid air, 1% NO and 2% NO_2 for up to 6 months at 100°C ~ 200°C, and iodine adsorption experiments on the aged adsorbents were conducted. Results show that humid air, 1% NO and 2% NO_2 have negative impacts on the performance of Ag^0Z and Ag^0-aerogel; especially, 2% NO_2 has the most severe impact on the iodine loading capacity of the both adsorbents. In addition, it was found that iodine loading capacity of the aged Ag^0Z decreased as aging time and temperatures were increased. Characterization studies were conducted with scanning electron microscopy (SEM), X-ray photoelectron spectroscopy (XPS), X-ray powder diffraction (XRD) and X-ray absorption fine structure (XAFS) to better understand the mechanisms of the aging processes on Ag^0Z and Ag^0-aerogel by observing the chemical and physical changes through aging processes. Kinetic models were developed and evaluated to simulate the aging processes in different gases; aging and iodine adsorption experiments on the Ag^0-aerogel are in progress. Reversible and irreversible reaction models were postulated to describe the reaction of the aging components with the silver adsorbent pellets. The kinetic studies of both Ag^0Z and Ag^0-aerogel will be discussed.

573577 Towards Inverse Design of Nanoporous Materials for Isotope Separation

The demand for isotopes has been increasing in recent years for applications especially to the fields of medicine, basic scientific research and energy production. Separation of isotopologues is technically challenging owing to their similarity in chemical properties. Traditional technology for isotope separation (e.g. cyclotron, chromatography, cryogenic distillation and etc.) has low capacity and often requires high energy consumption, leading to high operational costs. Nanoporous materials have shown a great potential in separation science, in particular for separating species with similar properties (e.g. N_2/CO_2). In this work, we explore applications of nanoporous materials either as adsorption media or membranes for separating a number of isotopologues. By combination of physics-based modeling and machine learning, we are able to identify best structural features (void fraction, largest included sphere diameter and sphere diameter along free sphere path) in achieving both high separation selectivity and capacity.

565907 Blind Source Separation in Raman and ATR-FTIR Spectroscopy: A Processing Case Study

Blind Source Separation in Raman and ATR-FTIR Spectroscopy: A Processing Case Study

Maggioni G.M., Kocevska S., Rousseau R. W., Grover M. A.

The low-level radioactive waste in the USA needs treating through vitrification to achieve long-term, safe, and environmentally sustainable storage [1].

To avoid the deterioration of the equipment, the process operating conditions must remain within a relatively narrow range of conditions. Ultimately, the key variable to monitor and control to achieve process safety and efficiency are the temperature, the species present in the feed, and their (relative) concentrations.

Spectroscopic techniques, such as Infrared (IR) (in the form of ATR-FTIR) and Raman have been long used to analyse and monitor the composition of solutions and slurries. The standard approach for the application of these techniques relies on calibration, usually a time-consuming procedure that needs to carefully design a set of experiments with known species and concentrations to estimate the model parameters.

However, each waste-containing vessel has not only been produced at different time and at different conditions, but it has also been long stored separately from the other vessels, thus experiencing different conditions and history. Additionally, if the species present in the mixture change, a new calibration typically becomes necessary, requiring to halt the process, lest one risks to obtain wrong results.

In the present work, we have developed a protocol to identify the spectra of the pure species and to compute a reliable estimate of their relative concentrations in a manner as calibration-free as possible, based on Blind Source Separation (BSS) techniques [2]. The protocol builds upon two different well-established techniques, namely Independent Component Analysis (ICA) and Multivariate Curve Resolution - Alternate Least Square (MCR-ALS).

In particular, we have investigated the application of our protocol on Raman and ATR-FTIR spectroscopy, using a simulant of the actual low-radioactive waste. We investigate the ability of BSS methods to correctly determine the spectra of unknown species in solution and discuss the reliability of one-point calibration methods to estimate the system composition.

[1] Civil Nuclear Waste Disposal, *Congressional Research Service*, 2018

[2] Hyvarinen, A. (2012). Independent component analysis: recent advances. *Philosophical Transactions of the Royal Society A* 371(1984), 20110534–20110534.

554686 Evaluation of Chemical Additives for Glycolate Destruction in High-Level Radioactive Waste Condensate in the Defense Waste Processing Facility

Savannah River National Laboratory (SRNL) has completed a work scope in support of the Savannah River Remediation (SRR) Nitric-Glycolic flowsheet. Glycolic acid will replace formic acid as a reducing agent in the pre-processing of High Level Waste sludge performed at the Defense Waste Processing Facility (DWPF). Glycolic acid will perform the same function as formic acid, namely reduction of mercury and adjustment of feed rheology and melter oxidation/reduction potential, while significantly reducing the potential for hydrogen generation in DWPF processing. Development testing has demonstrated glycolic acid virtually eliminates hydrogen production in the pre-processing steps.

The Nitric-Glycolic flowsheet significantly improves DWPF's ability to address Savannah River Site's key challenge - the incorporation of effluent received from the Salt Waste Processing Facility (SWPF). SWPF will deliver significant effluent volume to DWPF, resulting in a concurrent increase in DWPF effluent returned to the Concentration, Storage and Transfer Facilities (CSTF). This work scope demonstrates that glycolate can be destroyed under the Nitric-Glycolic flowsheet prior to returning the DWPF recycle stream to CSTF.

To avoid potential flammability issues due to thermolysis of glycolate in CSTF, SRR tasked SRNL to quantify and mitigate glycolate returns via DWPF recycle stream. The development of a strategy for glycolic mitigation initiated with a system's engineering workshop. Various chemical and/or physical solutions for how to and where to destroy glycolate were considered – consistent with DWPF operational capabilities and process requirements. The workshop identified chemical oxidation of glycolate within the DWPF Recycle Collection Tank (RCT)[1] as the most promising option with sodium permanganate and Fenton's reagent (metallic ion catalyzed hydrogen peroxide) as the two most promising oxidants.

SRNL between Fenton's reagent and permanganate could be made. A test matrix was generated to evaluate the impact of:

- Glycolate concentration,
- Oxidant type (sodium permanganate or Fenton's reagent),
- Oxidant strike time (rate of delivery),
- Oxidant stoichiometry,
- RCT solution temperature, and
- RCT solution pH.

Testing conditions were selected to reflect DWPF process capability and requirements. Non-prototypic concentrations of process organics were added to the process simulant to challenge the methodology and better support the down select. Tests were conducted with discrete liquid sampling events and online offgas analysis for data generation.

For the test conditions performed and data obtained, three conclusions are drawn.

- Sodium permanganate was clearly superior to Fenton's reagent as an oxidizing agent for destruction of glycolate in the RCT. Permanganate strikes consistently reduced glycolate concentrations from nominally 125-250 mg/L to below the target threshold of 50 mg/L. Corresponding strikes with Fenton's reagent did not successfully mitigate glycolate.
- Successful permanganate mitigations of glycolate were demonstrated across broad range of RCT process conditions. These conditions include the following.
    1. RCT solution starting pH: 3-13

2. RCT solution temperature: 15-50 °C
3. Oxidant strike time: 10 minutes to 2 hours
4. Glycolate concentration: 125-250 mg/L (with exploratory testing up to 2000 mg/L)
5. Oxidant addition amount: 100-150% stoichiometric requirement for all organics

SRNL, in conjunction with SRR, is continuing the glycolate mitigation development effort. Additional testing include protocol testing intended to clearly define strike parameters for engineering and operational implementation, off-normal testing for remediating foam-over conditions and testing at larger scale or with tank waste. Results from the testing will be discussed.

[1] The RCT collects offgas condensate from pre-processing, vitrification, and other unit operations performed in DWPF and is the singular return vessel delivering recycle effluent back to CSTF.

568856 Tank Closure Cesium Removal at Savannah River Site

High Level Waste (HLW) at the Savannah River Site (SRS) is currently stored in aging underground storage tanks. This waste is a complex mixture of insoluble solids (sludge) and soluble salts in an alkaline solution. Continued long-term storage of these radioactive wastes poses an environmental risk. The salt waste is currently treated to remove cesium by an interim solvent extraction process prior to disposal as grout in the Saltstone Disposal Facility. This interim process will be replaced with the Salt Waste Processing Facility (SWPF) to increase the salt waste treatment capacity. Treatment and disposal of the salt waste is the critical path to waste retrieval, risk reduction and closure of old style tanks at SRS.

Savannah River Remediation (SRR) is developing a supplemental at-tank ion exchange capability for the removal of cesium from liquid salt waste. Building on the conceptual development of the Small Column Ion Exchange (SCIX) process, the experience of modular commercial nuclear plant decontamination, and following the disaster response associated with Fukushima, SRR is deploying a modular at-tank treatment system for the selective removal of cesium from bulk salt waste effectively and efficiently. This Tank Closure Cesium Removal (TCCR) initiative will provide a supplemental treatment capability to existing and future salt processing and improve confidence in supporting the desired acceleration of waste retrieval and tank closure efforts.

The TCCR system configuration is consists of modular process structures located near Tank 10H and 11H. Salt waste will flow through a pre-filter to reject any solid material from entering the treatment system and be returned to Tank 10H. Clarified salt waste will then flow from the pre-filter into the first stage ion exchange column. The ion exchange media utilized will be highly selective for the removal of cesium and allow most other soluble constituents to pass through the media. Additional ion exchange columns are available and can be configured in series to improve decontamination factor, protect against flow "channeling" in the column, guard against media breakthrough, and allow for more continuous operation of the system. Decontaminated (cesium removed) effluent would flow from the ion exchange columns and sent to Tank 11H prior to being transferred to Saltstone for disposal.

Once the ion exchange media in a column becomes loaded with cesium to the extent practical ("spent"), the column with media will be removed from the system and replaced with a new ion exchange column loaded with fresh media. The spent column will be temporarily stored in an Interim Safe Storage (ISS) location within the SRS Tank Farm Facility. This ISS concept avoids the initial need for back-end process facilities (deferring the cost of these facilities) and allows for identification and evaluation of potential future disposal alternatives.

In addition to accelerating waste retrieval and tank closure activities, demonstration of this capability has several other benefits. Deployment of commercially available technologies with minor modification can expand Department of Energy (DOE) utilization of commercially supplied treatment processes. This "at-tank" treatment system could be modified to evaluate and demonstrate removal of other radioactive constituents (strontium, technetium, iodine, etc.) of bulk waste. The modular aspect of this capability can allow for additional units to be delivered and deployed at SRS, as well as, provide supplemental treatment capabilities for waste treatment at other DOE sites.

577672 Large Cemented Gibbsite Agglomerates in Alkaline Nuclear Waste at the Hanford Site and the Impacts to Remediation

Jet-sluicing has been the primary method of moving nuclear waste sludge between storage tanks at sites such as Hanford in the United States. Recent experience at an area of Hanford comprising twelve, 530,000-gallon underground tanks (C Farm), with waste that had aged in place for decades, showed that jet sluicing left a small heel of waste in the tanks. This study sought to determine why this material was left behind, focusing on Hanford tank 241-C-109. The waste heel in this tank was sampled and then analyzed by X-ray diffraction and scanning electron microscopy (SEM) with energy dispersive spectroscopy. Large gibbsite [$Al(OH)_3$] particles were found to be the major component in the waste heel. Individual, primary gibbsite particles were to 20 to 100 microns in diameter but were cemented together to form agglomerates as big as seven centimeters in diameter that were difficult to crush with a laboratory mortar and pestle. SEM showed that either amorphous aluminum hydroxide or gibbsite itself was the cementing agent. While large cemented gibbsite nodules have been reported in acid soils in nature, this study shows that similarly large gibbsite agglomerates can form in alkaline environments when aged for sufficiently long time, here for over 40 years prior to waste retrieval. Fluid dynamics calculations indicate that these cemented particles would be difficult to mobilize with standard jet slurry technologies. Even if successfully mobilized, the particles would be difficult to keep suspended and would not fit into transfer lines that are commonly 5-cm pipe diameter. Thus, these large gibbsite particles explain some of the difficulty in suspending aged alkaline nuclear waste with standard sluicing technologies.

571866 Processing of Actinide Materials for Reuse/Disposal

The US Department of Energy (DOE) manages materials that contain a range of actinide isotopes that were produced from the 1960s through the 1980s by irradiating targets in production reactors to produce heavy isotopes and enriching them for DOE programmatic use, scientific research, and industrial and medical applications. Because the production reactors and enrichment facilities have been shut down, many of these unique materials will never be produced again. Oak Ridge National Laboratory (ORNL) enriched many of the materials and has since used them in DOE's center for production, storage, and distribution of heavy isotopes (uranium through californium) for the heavy-element research program. As a result, ORNL has accumulated an inventory of rare actinide isotopes that are being held for reuse because they have potential intrinsic value to DOE, even though the materials have no currently defined use. Steps are being taken to evaluate the inventory, recover the useful materials, and distribute them to programmatic users. If no programmatic need for the materials is found, the materials are processed for transuranic waste disposal. This paper describes the processes that are being used to recover or disposition the materials.

563598 In-Situ Raman and ATR-FTIR Spectroscopy for Monitoring of Molecular Species in Low-Activity Nuclear Waste

A significant portion of the nuclear waste stored at the Hanford site is low-activity waste, primarily made up of non-radioactive components such as sodium salts. The DOE has initiated the Direct Feed Low-Activity Waste (DFLAW) project to expedite the clean-up of low-activity waste. Under DFLAW, a portion of the liquid low-activity waste will be preprocessed to reduce cesium levels and solid content and directly sent for vitrification. The complexity of the waste and its hazardous nature make the analysis challenging, thus reliable measurements of the composition and solid content are needed to facilitate processing.

In our present research, we are exploring use of in-situ monitoring tools, such as ATR-FTIR and Raman spectroscopy, to collect high-dimensional measurements of the waste. The data is collected on simulated waste mixtures that represent different DFLAW locations at Hanford. Our approach includes collecting training sets for the simulated mixtures that are used to build regression models. Next, we use the models to predict the composition of unknown mixtures. Our goal is to use the experimental findings to compare the performance of Raman and ATR-FTIR probes and build a robust model that will generate the composition of the multicomponent waste system.

Our research has shown that using in-situ ATR-FTIR and Raman spectroscopy, coupled with data-driven models can be used to monitor key species in simulated nuclear waste mixtures. This approach has the potential to reduce or eliminate the need for offline sample collection, which will minimize employee exposure to hazardous waste, while improving sample turn-around times.

558836 Ultra-Microporous Materials for Efficient Removal of Krypton from Nuclear Reprocessing Facilities

Reprocessing of used nuclear fuel (UNF) is a prerequisite for any future expansion of civilian nuclear power. Major challenges during the reprocessing of nucler fuel is the release of volatile radio nuclides that must be captured and subsequently stored for an extended period of time before being released into atmosphere. While methods such as cryogenic distillation and fluorocarbon based absorption have been proposed and tested, solid state adsorbents in particular permanently porous Metal-Organic Frameworks (MOFs) are considered better alternatives in terms of cost and engineering control to capture these volatile radionuclide including ^{85}Kr ($t1/2 = 10.8$ years). However, because of presence of more polarizable Xe during the reprocessing, adsorbent materials including MOFs are found to be more Xe- philic than Kr. To address this challenge, a two-bed breakthrough method is being developed herein which leads to enhanced Kr adsorption and separation of ^{85}Kr during nuclear reprocessing. The extent of enhancement of Kr adsorption depends on the nature of adsorbing materials. Here, we evaluate the Kr adsorption capability of a series of hybrid ultra-microporous materials, namely SIFSIX-3-M (M= Zn, Cu, Ni, Co or Fe) for Kr adsorption and separation performance. The Fe-analogue in particular shows excellent Kr adsorption than the current benchmark materials: CC3, Ag-mordenite, NiMOF-74 and SBMOF-1, under simulated nuclear reprocessing off-gas conditions. The higher Kr adsorption and separation efficacy was attributed to the precise pore diameter of the Fe-analogue for such separation. All materials were systematically studied for gamma irradiation stability, which reveals that the metal center in these isostructural materials plays a crucial role in their irradiation stability. Zn and Ni analogues show phase change at 1KGy, while SIFSIX-3-Fe shows phase change at 3KGy. SIFSIX-3-Co structure was stable up to 10KGy before it undergoes phase change while SIFSIX-3-Cu was the most stable analogue. The phase change of all materials were interpreted using PXRD. These results attest the potential of these materials for Kr removal under the nuclear reprocessing conditions.

567054 The Migration Law of Nuclides in Steam Reforming Reaction of Spent Radioactive Resins

The treatment of spent radioactive resins in the nuclear industry is important due to its great potential hazard to human beings and environments. The two keys during the disposal of the resins are volume reduction ratio and leaching resistance. Considering these two aspects, the steam reforming technology is a better method for resin disposal with high volume reduction ratio (>6) and radionuclides retention efficiency (>99.99%). During the reaction, nuclides are released when the spent resins are pyrolyzed into light hydrocarbons, sulfides, carbon oxides and a little residue at 650-750°C. Meanwhile, the nuclides are contacted with silicon-aluminum compounds and structure directing agents in the reactor to form specific mineral, which can immobilize the nuclides in its structure. The key to the reaction is that the nuclides are released efficiently from the spent resins with large volume, and then immobilized by minerals with small volume. Therefore, it is of great significance to study the migration law of nuclides to regulate the immobilization efficiency. In this work, the migration process of nuclides in the pyrolysis of resins under the steam reforming reaction conditions is investigated in the fixed bed reactor. The pyrolysis products obtained from different reaction times and temperatures are analyzed by SEM-EDX to observe the morphologies and nuclides distribution. The form of nuclides are studied by XRD, and the quantification of nuclides are obtained by ICP-MS. It is found that different migration paths of nuclides occur in different pyrolysis stages of resins, and a migration model is proposed to describe the phenomenon. Most of the nuclides are retained in the residue with the form of oxides and inorganic salts.

Keywords: Migration law; Nuclides; Steam reforming reaction

575764 Design and Operation of a System for the Preparation of a Molten Chloride Nuclear Fuel Salt

The design and operation of a system to prepare a uranium containing chloride fuel salt from impure components will be presented. Impure NaCl and CaCl2 are first dehydrated using a combination of vacuum heating and reaction with HCl gas. Metal impurities are removed using Ca metal. Uranium is oxidized into the salt using FeCl2 as an oxidant to form UCl3. The success of the process is determined in-situ by analyzing the salt using cyclic voltammetry and analyzing the off-gas using Quadrupole Mass Spectrometer or acid-base titration. The purity of the salt will be confirmed using TGA/DSC and ICP-MS. The design considerations of the reaction system will be discussed including the quartz and teflon reaction vessel. The prepared chloride fuel salt may be used to conduct neutron experiments in a test reactor such as the planned Versatile Test Reactor.

558835 Neutron Radiography for the Determination of Molten Chloride Viscosity, Density, and Homogeneity

Introduction:

Advanced molten salt reactor (MSR) designs utilize actinide fuel dissolved in molten salt, which results in a high temperature liquid fuel solution. The molten salts typically used are mixtures of alkali metal halides and/or alkaline earth metal halides, conventionally fluorides.[1] Recent findings have identified chloride salts as a promising alternative, and the nuclear, chemical, and physical properties of chloride salts and chloride salt mixtures are needed to advance molten salt reactor design.[2] Solutions of $NaCl/PuCl_3/UCl_x$ are emerging as the most viable candidates for liquid fuel. However, limited thermophysical property data exists in the literature for such solutions, and these data as a function of composition and temperature are essential inputs to modeling the performance of this fuel form.

Neutron radiography is a well-established technique for the nondestructive visualization of the interior of complex systems.[3-5] In neutron radiography, the attenuation of a parallel neutron beam due to absorption and scattering of the neutrons by the sample is measured with a position-sensitive detector. This method can be extended by the use of time-resolving detectors to real-time observations and is then called "dynamic neutron radiography". Coupling dynamic neutron radiography with conventional techniques (e.g., falling sphere viscometry) allows physical property measurements to be made of samples in extreme environments (e.g. molten salts at 850 °C). This combination of techniques has the potential to produce property data with more accuracy and precision than any previous molten salt measurement. Furthermore, this approach is ideal for such systems because it is performed in-situ, allows for a wide range of experimental conditions, and is able to detect and account for aberrations in density and concentration.

The goal our project is to develop and demonstrate the ability of neutron radiographic techniques to determine the viscosity, density, and homogeneity of molten chlorides as a function of temperature. Experiments will focus on non-radiological and low-level radiological salt systems of interest to both the MSR <u>and</u> pyroprocessing communities. The results from these trials will lay essential groundwork for subsequent experiments, which will integrate plutonium-bearing salt into the test matrix.

Experiment:

Kahle, Winkler, et. al. successfully demonstrated high-temperature (2300 K) dynamic neutron radiography with the falling sphere method at SACLAY in 2003.[6] In that work, physical properties of silicate melts were measured using a graphite tube furnace, which allowed for in-situ and real-time observations of large melt volumes. From these observations, the velocities (v) of two falling spheres of known density were determined. Two spheres were used so that the determination of both the melt viscosity (η) and melt densities (ρ_{melt}) could be calculated using Stokes' law.

Our experiments will be modeled after this work. It will be conducted in a sealed stainless steel vessel with the molten chloride solution contained within a ceramic liner, into which a witness K-type thermocouple will be inserted from the top of the apparatus. bbjects of a range of known densities and geometries (e.g., spheres, bullets) will be loaded into a hopper at the top of the apparatus, and will be dropped individually by an electronically actuated gate. The assembly will be heated externally by a clam-shell furnace, which will be surrounded by alumina insulation. The field of view of the neutron radiography will center on the molten chloride solution.

Utilizing the pixelated MCP detector available on Flight-Path-05/ERNI to record 2-D neutron radiographs of the borated metallic objects falling through the molten chloride solution, accurate determinations of solution density and viscosity can be achieved. This type of measurement will be then repeated over a range of temperatures and for various solution compositions, thus allowing us to map the density and viscosity of the molten chloride solutions as a function of temperature and composition. Additionally, using neutron pair function distribution analysis, homogeneity will be determined by examining liquid-liquid phase transitions. Exploiting the time-of-flight information recorded from the MCP detector, neutron transmission spectra will be recorded on a pixel-by-pixel basis. By fitting each pixel-based transmission spectra with the R-matrix code SAMMY, isotopic specific information can be determined for each pixel, thus providing valuable information on the degree of mixing throughout the solutions.

Beam time experiments will be carried out in July and September 2019. Results from these experiments will be discussed at the conference

References:

1. Haubenreich, P.N. and Engel, J.R. "Experience with the molten-salt reactor experiment." *Appl. Technol.* 8, 118 (**1970**).
2. Holcomb, D.E.; Flanagan, G.F.; Patton, B.W.; Gehin, J.C.; Howard, R.L.; Harrison, T.J. "Fast spectrum molten salt reactor options." ORNL/TM-2011/105 (**2011**).
3. Lehmann, E.; Pleinert, H.; Williams, T.; Pralong, C. "Application of new radiation detection techniques at the Paul Scherrer Institut, especially at the spallation neutron source." *Nucl. Instrum. Methods in Phys. Res. A* 424, 158 (**1999**).
4. Winkler, B.; Knorr, K.; Kahle, A.; Vontobel, P.; Lehmann, E.; Hennion, B.; Bayon, G. "Neutron Imaging and neutron tomography as non-destructive tools to study bulk-rock samples." *J. Mineral.* 14, 349 (**2002**).
5. Tremsin, A.S.; Perrodin, D.; Losko, A.S.; Vogel, S.C.; Bourke, M.A.M.; Bizarri, G.A.; Bourret, E.D. "Real-time Crystal Growth Visualization and Quantification by Energy-Resolved Neutron Imaging." *Rep.* 7, 46275 (**2017**).
6. Kahle, A.; Winkler, B.; Hennion, B.; Boutrouille, P. "High-temperature furnace for dynamic neutron radiography." *Sci. Instrum.* 74, (8), 3717-3721 (**2003**).

LA-UR-19-22600

561069 Electrochemical Corrosion Measurements in Molten Flinak Salt

Molten fluoride salts have attracted interest as fuel and/or coolant for several different molten salt reactor (MSR) designs. Fluoride salts have many advantageous properties for these applications, such as high ionic conductivity, good heat transfer capacity, radiation insensitivity and high solubility for actinides. However, the corrosion of storage materials owing to impurities in the molten salt has not been thoroughly studied, especially in fluoride salts. Some researchers have studied corrosion in fluoride salts, but they only compared the morphologies and the elemental composition of the metal's surface or near surface region after static corrosion tests. LiF-BeF_2 (FLiBe) and LiF-NaF-KF (FLiNaK) are candidate fluoride salts for these applications. In this study, we used two electrochemical methods for measuring corrosion rate in FLiNaK: linear polarization and zero resistance ammeter (ZRA). ZRA is the most suitable technique for monitoring corrosion processes in situ, since it yields real time data that can be directly related to corrosion without applied potential. ZRA data will be presented and compared to morphologies and element mapping of the surface of the metal alloys obtained by using scanning electron microscopy (SEM) and energy dispersive spectroscopy (EDS). In addition, the composition of the salts was analyzed by inductively coupled plasma – mass spectroscopy (ICP-MS) before and after ZRA measurements to determine which corrosion products are soluble in the molten fluoride salt.

578240 Thermodynamic Analysis of Minor Components in Molten Salt Mixtures

This study focuses on the various models that can be applied to predict the thermodynamic behavior of minor components in molten salt mixtures, defined here as any component with a mole fraction of approximately 0.02 or less. Defined in this way, minor components include corrosion products in molten salt mixtures in general, and fission products in liquid-fueled molten salt reactors, making them important for corrosion control and for separations processes used in nuclear reprocessing. Various factors that affect their equilibrium behavior, especially concentration and base salt composition, were examined using published experimental data from both older and recent studies.

A common reference state for activity in current molten salt literature is that of the pure component at the given temperature in either a solid or liquid state. The liquid has been particularly recommended as being closer to the physical state of the component since the molten salt mixture is itself a liquid. A minor component in a molten salt mixture interacts predominately with the base salt components rather than with itself. This means that use of a pure component reference state requires activity coefficients to make large corrections in order to accurately represent interactions with a given salt. The activity coefficient is also the only way to quantify the influence of different base salts. The present study examines an alternative model that encapsulates the important interactions between minor components and the base salt mixture in a single number. An advantage of this method is that it requires only slight activity coefficient adjustments over the concentration range of interest when estimating the reduction potential of a minor component. This can result in significantly less error than would otherwise be the case—0.02 V or less instead of 0.1 V or more. Use of the method to improve the accuracy of Pourbaix diagram-style corrosion analysis of molten salt mixtures is demonstrated.

560753 Metal Dissolution in Molten Salts for Redox Control

Abstract

Molten chloride salts have great potential for higher efficiency heat transfer compared to nitrate-based salts. The chloride-based salts can operate at higher temperatures and in turn produce more electricity. The redox potential of the chloride salts is much greater at higher temperatures and can easily corrode common alloys. In this study, the electrochemical effect of adding magnesium metal to a $MgCl_2$-KCl-NaCl salt mixture was investigated. The solubility of the magnesium in the $MgCl_2$-KCl-NaCl salt mixture was analyzed at different temperatures to determine the maximum saturation point. Through open circuit potentiometry it can be seen that the redox potential of the salt decreased when magnesium was added. Cyclic voltammetry shows both a reduction and oxidation peak, with the reduction peak being greater in size. These results signify that the addition magnesium metal could be an advantageous medium to control redox potential in the $MgCl_2$-KCl-NaCl salt.

572105 Habit 2.1: Assessing Consequences of Chemical Releases on Control Room Habitability Considering the Impact of a Dividing Streamline

HABIT is a package of computer codes designed to assist in the evaluation of control room habitability in the event of accidental spills of hazardous materials in the area and is part of the U.S. Nuclear Regulatory Agency's Radiological Protection Computer Code Analysis and Maintenance Program (RAMP). Developed at Pacific Northwest laboratory (PNL), HABIT was originally written to model atmospheric dispersion based on standard passive dispersion models. In addition to an extensive update of the user interface, HABIT 2.0 was also extended to include the effects of denser-than-air contaminant cloud behavior using the SLAB and DEGADIS models. This paper summarizes how the previous HABIT source models (instantaneous and finite duration releases) were used to generate inputs to SLAB and DEGADIS. HABIT 2.1 has considered how the concept of a dividing stream line can successfully refine the predicted concentration expected at a control room air intake. Model results are presented for a hypothetical chlorine release which reflects the importance of denser-than-air behavior on the potential impact of the air intake height on control rooms of light-water reactors (LWR).

576103 Natural Circulation in a VERY High Temperature Reactor (VHTR)

The generation-IV nuclear reactors, Very High Temperature gas-cooled Reactors (VHTRs), are proposed by the US Department of Energy to produce Green House Gas emission-free electricity and hydrogen. VHTRs are designed with passive safety features to withstand hypothetical accidental scenarios such as loss of coolant accidents, earth quakes and tsunamis, etc. But it is not well understood how the VHTR's passive safety designs will react to the aforementioned accident scenarios. Knowledge about the natural circulation flow and heat transfer in Pressurized Conduction Cooldown (PCC) and Depressurized Conduction Cooldown (DCC), and the effects of bypass flow in forced circulation flows are essential for safety of VHTRs. Further, an air ingress scenario in which ambient air would enter the core through a break in the coolant flow loop piping could cause oxidation of a hot graphite core and severely damage the core. In the present study, (1) the natural circulation flow and heat transfer, (2) the effects of air ingress at different pressures and temperatures (3) forced convection and the effect of bypass-coolant flow phenomenon have been investigated experimentally.

To understand the natural circulation flow phenomena in a VHTR and study the effects of air-ingress, natural circulation experiments have been performed by injecting nitrogen (instead of air to avoid graphite oxidation) into the lower plenum of a helium-filled flow loop consisting of a riser and downcomer connected by a lower plenum at the bottom and upper plenum at the top. The riser was made of a graphite test section containing a single flow channel and electrical heater rods and instrumented with thermocouples. The transport of nitrogen through the graphite flow channel has been quantified by measuring the volumetric concentrations of nitrogen in lower and upper plena. The transport of nitrogen through the graphite test section could be determined by detecting small but sudden changes in the local graphite temperatures at different axial elevations. The steady state gas circulation rates of the helium-nitrogen mixture between the hot riser and cold downcomer have also been measured. The experimental findings indicate that the driving mechanism for nitrogen transport in the air-ingress scenario results from both molecular diffusion and natural circulation. At low graphite temperatures in the riser, molecular diffusion is the dominant mechanism for nitrogen transport, but as the graphite temperature increases, the natural circulation becomes more dominant, and the nitrogen transport rate increases.

578158 Speciation of Fuel and Fission Products in Molten Salt Solutions from Ab Initio Molecular Dynamics

Impurities including fuel, fission products, and corrosion products are formed and consumed within molten salt reactors. These dissolved impurities will likely affect salt structure and thermophysical properties as they interact with each other and the solvent environment, and understanding these effects is crucial to the safe and effective design of reactors and the economical production of clean energy. We use first principles molecular dynamics to examine solute-solute interactions within molten fluoride and chloride salts. Simulated density and structure are compared to experimentally-determined values for validation. Our simulations find that solute-solute interactions have a significant effect on solute speciation, with—for example—uranium forming diverse complexes with multiple formal oxidation states at process-relevant concentrations. Molybdenum, another solute expected to be generated in the molten salt as both a fission and corrosion product, is found to form polynuclear molybdenum halide anions in solution, with possibly profound implications for the stability of molybdenum in the salt. Water is introduced, representing contamination of the salt, and is found to have significant effects on anion stability. Extensive charge analysis is used to analyze solute-solute and solute-solvent interactions. The applications of these results to thermophysical property prediction and fission product separation is discussed.

558099 Density Functional Theory Study of the Tritium Diffusion, Desorption, and Trapping on the γ-$LiAlO_2$ (100) Surface

γ-$LiAlO_2$ enriched with the ^6Li isotope is one of the important components of tritium-producing burnable absorber rods (TPBARs). This oxide is located between the zircaloy-4 liner and nickel-plated zircaloy-4 tritium getter. When irradiated in a pressurized water reactor (PWR), the ^6Li pellets absorb neutrons, simulating the nuclear characteristics of a burnable absorber rod, and produce tritium ( ). The specie chemically reacts with the metal getter, leading to the formation of a metal hydride. Currently, the understanding of ^3H transport through the ceramic pellets and the barrier/cladding system is hampered by the lack of fundamental data particularly related to hydrogen isotope solubility and diffusivity.[1] Recently, by employing first-principles density functional theory (DFT) calculations, we have investigated the energetics of the charged defects (including vacancies, interstitials, antisite defects, and ^3H substitutional defects) of the γ-$LiAlO_2$ used in TPBARs and identified the trends in stability properties of various point defects under TPBAR operating temperature and O_2 partial pressure conditions, and the tritium diffusion pathways in γ-$LiAlO_2$ in the presence of interstitial and substitution AL Li defects, hydroxide (O-^{3}H) vacancy defect, Al (or Zr) defect, and of the interactions of ^3H with O-vacancies.[2-9]

Once the is produced in the bulk of γ-$LiAlO_2$, it diffuses to its surface. It is possible that atoms bind with surface atoms or become trapped at vacancy sites. With increasing the number of diffused from bulk to surface, upon recombination and desorption from surface sites, they will escape from the γ-$LiAlO_2$ surface to form different species (possibly being T, OT, T_2, or T_2O, which need to be identified) and are captured by the getter. Hence, exploring the diffusion from bulk to surface, corroborated with surface adsorption, diffusion and recombination and followed by desorption from surface as various types of species becomes a very important task as an integral part of the solubility and diffusivity in the pellets. Based on DFT, in this study, we explored the diffusion, desorption, and trapping on the most stable γ-$LiAlO_2$ (100) surface. Firstly, we investigated the diffusion to the nearest neighbor (NN), 2nd NN and 3rd NN Li vacancy (V_{Li}) and the TO diffusion to a bound O and Li vacancy (V_{LiO}). The results showed that the energy barriers of T diffusion to the NN, 2nd NN, and 3rd NN V_{Li} (1.40, 2.05 and 2.93 eV) on (100) surface are larger than those of T diffusion in the bulk (0.63, 1.51, and 0.86 eV).[2] The energy barriers of TO diffusion on the surface to the NN, 2nd NN, and 3rd NN (1.95, 2.25 and 2.96 eV) V_{Li} are comparable with those of T diffusion on the same surface, but are much smaller than those of TO diffusion in the bulk (2.17, 4.23, and 3.44 eV).[2] Secondly, we considered these T/TO diffusion on the defective (100) surface under irradiation from MD simulation. By analogy with 2nd NN and 3rd NN V_{Li} on the corresponding perfect surface, the energy barriers for T and TO diffusion cross the hole to 2nd NN and 3rd NN V_{Li} are 4.74, 4.85 and 1.78, 2.28 eV respectively. Therefore, the order of energy barrier for substitution T diffusion in bulk and surfaces is: bulk < surface < defective surface, which is reversed for TO diffusion, because the space is an advantage for TO but is a disadvantage for T diffusion. Furthermore, we explored the diffusion from bulk to the (100) surface with/without V_{Li}. The obtained energy barrier for diffusion from bulk to (100) surface with/without V_{Li} is 1.37/1.59 eV, and for T to desorb from the surface is 4.36/3.10 eV with/without V_{Li} respectively. Thus, it is easy for to diffuse from bulk to (100) surface but is difficult for T to desorb from the surface. Finally, we considered the trapping in the surface V_{Li}. Our results indicated that one V_{Li} can trap two , and two V_{Li} can trap a maximum of three . More T could be adsorbed on the nearby Li/Al top site, and even combined into T_2 or T_2O molecule. The T_2 molecule can desorb from surface spontaneously while the T_2O desorption needs energy of 0.69 eV. Therefore, the T_2 and T_2O molecules are main products from (100) surface of γ-$LiAlO_2$. Since the T_2O is formed by two T with a surface O which only has one coordination associated with two V_{Li}, increasing V_{Li} can bring out more such O to produce more T_2O.

Reference:

1. Senor, D. J. *Recommendations for Science and Technology in Support of the Tritium Sustainment Program*, PNNL-27216; PNNL: 2017.
2. Paudel, H.; Lee, Y. L.; Senor, D.; Duan, Y., *Tritium Diffusion Pathways in γ-$LiAlO_2$ Pellets Used in Tpbar: A First-Principles Density Functional Theory Investigation*. J. Phys. Chem. C **2018**, *122*, 9755-9765
3. Paudel, H.; Lee, Y.-L.; Holber, J.; Sorescu, D. C.; Duan, Y. *Fundamental Studies of Tritium Solubility and Diffusivity in $LiAlO_2$ and Lithium Zirconates Pellets Used in TPBAR*; DOE/NETL-PUB-21464, doi: 10.2172/1463897, 2017.
4. Paudel, H. P.; Duan, Y., A First-Principles *Density Function Theory Study of Tritium Diffusion in Li2zro3: Application for Producing Tritium*. J. Phys. Chem. C **2018**, *122*, 28447-28459.
5. Lee, Y. L.; Holber, J.; Paudel, H.; Sorescu, D. C.; Sensor, D.; Duan, Y., *Density Functional Theory Study of the Point Defect Properties for the γ-$LiAlO_2$, Li_2ZrO_3, and Li_2TiO_3 Materials*. J. Nucl. Mater. **2018**, *511*, 375-389.
6. Jia, T.; Zeng, Z.; Paudel, H. P.; Senor, D. J.; Duan, Y., *First Principles Study of the Surface Properties of Γ-Lialo2: Stability and Tritium Adsorption*. J. Nucl. Mater. **2019**, submitted.
7. Duan, Y.; Sorescu, D. C.; Senor, D. J., *Theoretical Investigation of the Electronic Structural, Optical, Thermodynamic, and Thermo-Conductive Properties of γ-$LiAlO_2$ with ^6Li Isotope Substitutions for Tritium Production and CO_2 Capture*. J. Solid State Chem. **2019**, submitted.
8. Paudel, H. P.; Lee, Y.-L.; Fedele, C.; Duan, Y. *Fundamental Studies of Tritium Diffusivity in Irradiation Defective Lithium Aluminate and Lithium Zirconate Pellets*; DOE/NETL-PUB-22212, 2018
9. Jia, T.; Paudel, H. P.; Duan, Y. *Fundamental Studies of Tritium Diffusivity and Formation on the Surfaces of Defective $LiAlO_2$ Pellets*; DOE/NETL-PUB-22213, 2018.

574009 Fundamental Studies of Tritium Diffusivity in Carbon Doped Defective γ-$LiAlO_2$ Pellets

Fundamental Studies of Tritium Diffusivity in Carbon Doped Defective γ-$LiAlO_2$ Pellets

Hari P. Paudel[1], David J. Senor[2], Yuhua Duan[1]

[1] National Energy Technology Laboratory, United States Department of Energy, Pittsburgh, Pennsylvania 15236, USA

[2] Pacific Northwest National Laboratory, Richland, Washington 99354, USA

Abstract:

γ-$LiAlO_2$ is a ceramic material with superior thermo-physical and thermo-chemical properties, and is highly compatible with other materials being used in the form of annular pellets in tritium-producing burnable absorber rods (TPBARs) to produce tritium (^3T) by thermal neutron irradiation of ^6Li. In addition to that γ-$LiAlO_2$ has an excellent irradiation behavior at high temperature, and is more swelling resistant than many other Li rich materials[1,2]. The ^3H diffusion and recovery processes have been studied experimentally in ceramic breeding materials irradiated with neutron beams by Okuno and Kudo[3,4]. It was revealed that the tritium produced in the ceramic materials (Li_2O, γ-$LiAlO_2$, Li_2SiO_3, Li_4SiO_4, Li_2ZrO_3, Li_8ZrO_6) irradiated with neutron beams was released mainly in the chemical form of tritiated water (HTO) when heated in vacuum[4]. The step before release of ^3H to gas gap in reactor in the form of HTO involves an atomic migration of ^3H through grains of radiation damaged material in bulk.

The study of Li diffusivity in Li containing ceramics such as γ-$LiAlO_2$ and Li_2ZrO_3 has also been a subject of interest in Li-ion battery and solid oxide fuel cells. In the past, by using NMR and dc conductivity measurements in γ- $LiAlO_2$, it was found that the Li diffusion coefficient is in the range 10^{-20} to 10^{-13} m^2/s in the temperature range of 400 K to 1000 K[5]. In the same experiment, the Li activation energy was found to vary between 0.74 to 1.14 eV. By calculating several different diffusion pathways, Paudel *et al* show that the smallest activation energy barrier for ^3T in pure γ-$LiAlO_2$ is 0.63 eV for substitutional tritium diffusion and corresponding diffusion coefficient is 3.25×10^{-12} m^2/s[6].

The experimental observation in a polycrystalline γ-$LiAlO_2$ irradiated with both $^4He^+$ and $^1H^+$ ions reveals that there is a formation of isolated elementary voids along with other phases of lithium and aluminate (LiH and $LiAl_5O_8$)[7]. These elementary voids are believed to trap diffusing entities (Li and ^3T) and hinder their diffusivities at the elevated temperatures. Identifying trapping sites of light elements like Li and ^3T requires image resolution in atomic scale, and so far it is not clearly understood experimentally. Here we present theoretical studies based on the first principles density functional theory to calculate the formation of voids and their influences on the diffusion barriers of ^3T. In particular, we calculate different pathways of ^3T migration in presence of voids and find a minimum energy barrier for diffusion in bulk. We also examine vacancy-vacancy interaction energy to understand the properties of voids., We find that the interaction
st

energy is 0.12 eV between two 1 NN V_{Li}. It decreases monotonically as distance between them increases. In experiment, it has also been reported that there is presence of carbon impurities in defective γ-$LiAlO_2$. We calculate the formation of C-impurity and the diffusion barrier in C-doped γ-$LiAlO_2$, and compare the barrier energies with the case of pure γ-$LiAlO_2$. Our calculation shows that the smallest diffusion barrier for ^3T is found to be 0.98 eV in a C-doped γ-$LiAlO_2$ which is 0.35 eV higher than that in pure γ-$LiAlO_2$.

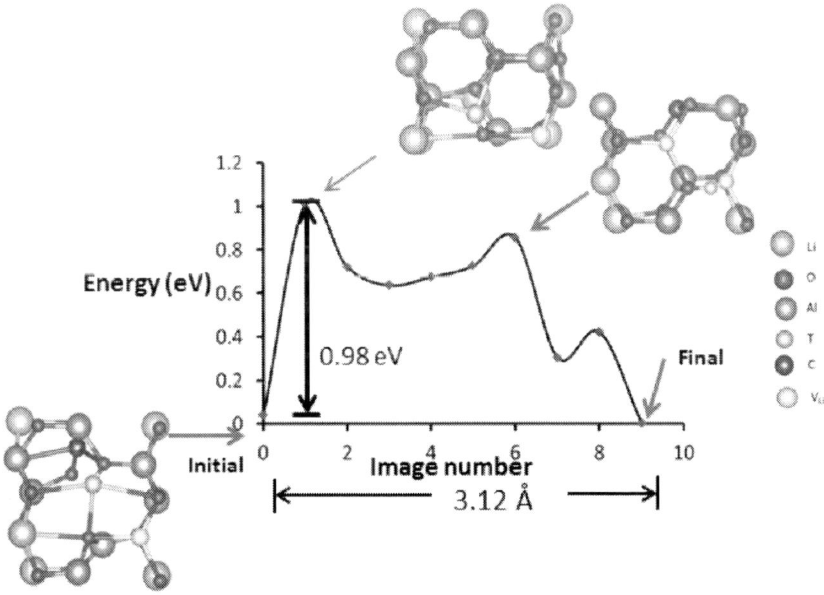

Figure: Substitutional Tritium diffusion in presence of interestitial C impurity.

References

Cao, H. *et al.*, *Solid State Ionics* **2005**, *176*, 911-914

Li, L. J. *et al*, *J. Mater. Chem. A* **2015**, *3*, 894-904

Okuno K. *et al.*, *J. Nucl. Mater.* **2001**, *299*, 242-249

Okuno K. *et al.*, *Fusion Eng. Des.* **1989**, *8*, 355-358

Indris, S. *et al.*, *J. Phys. Chem. C* **2012**, *116*, 14243-14247

Paudel, H. P. *et al.*, *J. Phys. Chem.* C **2018,** *122*, 9755

Jiang, W. *et al.*, J. Nucl. Mater. **2018,** *511*, 1

MODELING OF PLUGGAGE AND SPATIAL DISTRIBUTION OF PARTICULATE MATTER DEPOSITED IN THE SAVANNAH RIVER SITE H-CANYON PROCESS VESSEL VENT FILTERS

Dr. James E. Laurinat, Savannah River National Laboratory, Aiken, SC

Dr. Michael G. Bronikowski, Savannah River National Laboratory, Aiken, SC

Summary and Introduction

A model was developed to predict the rate of pluggage and spatial distribution of fissile particulate material in the Savannah River Site H-Canyon Process Vessel Vent (PVV) filters. The PVV filters are multilayer fiberglass mat filters in the off-gas vents from H-Canyon. Exhaust fans pull H-Canyon air through these filters, which are located upstream from the final sand filter and the exhaust stack. During operation of the Plutonium Uranium Extraction (PUREX) process in H-Canyon, the PVV filters periodically become plugged with ammonium nitrate. When pluggage occurs, the vent flow is shut down, and the PVV filter are flushed. Fissile material that accumulates in these filters between flushes poses a nuclear criticality risk. A study was requested to evaluate the feasibility of extending the interval between flushes and the possibility of raising the criticality safety limit.

Description of Vessel Vent Filters

There are separate PVV filters for the hot (high radioactivity) and warm (moderate radioactivity) sides of H-Canyon. The vent air from the hot side of the canyon flows through Filter 7.2F, and the vent air from the warm side of the canyon flows through Filter 5.7F. Each of these filters is 4.75 m in height and 2.44 m in diameter. Each PVV filter is divided into two separate sealed filter sections, an upper and a lower section. The filter assembly within each section is 1.397 m high. Vented air flows into and out of the PVV filter through 0.305-m diameter inlet and outlet pipes on the top of the filter. Upon entering the PVV filter, the vented air flow divides equally between the top and the bottom filter sections. The air flows from the bottom of each section through fiberglass filtering media and out the top of the section as shown in Figure 1. The air is filtered through five fiberglass layers separated by 4-mesh stainless steel screens held in place by a stainless steel frame 0.0445 m thick. The first four layers are made of 115K fiberglass packed to varying heights and densities, as indicated by Figure 1. The final layer is made of PF 105 AA fiberglass.

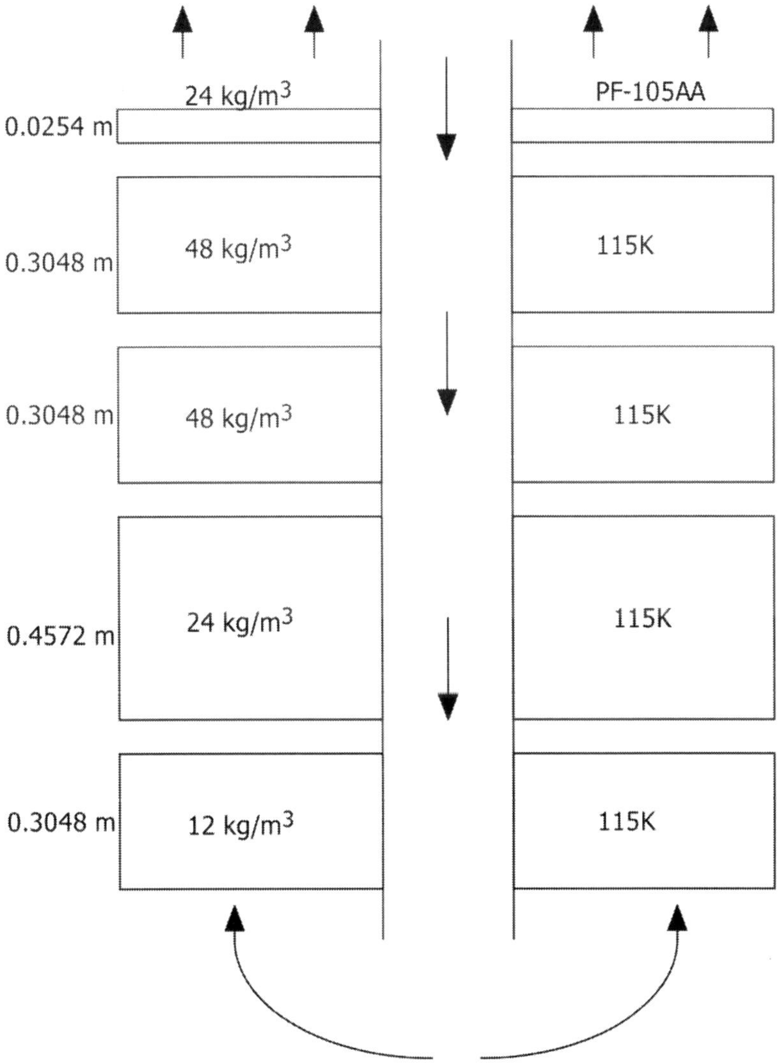

Figure 1. PVV Filter Bottom Section Schematic Showing Filter Thickness, Filter Fiber Density and Type, and Air Flow Rate through the Filter

The filtration efficiency of each filter layer is determined by the properties of the fiberglass fiber material and the manner by which the filter is loaded. The PF 105 AA (MIL B5924B Type 1) was purchased from HITCO (now TIGHITCO). The PF 105 AA specifications are: a thickness of 0.0109 – 0.0147 m, a density of 8.97 – 10.25 kg/m^3, a binder content of 15-20%, and an average fiber diameter of 1.0 $\pm$ 0.25 μm. Five layers of this half inch fiberglass are baked to remove the binder and are compressed from 0.038 m to 0.032 m to use as the final filter. Since 1984, only a 0.0254-m final filter thickness has been used. The 115K fiberglass (a curly-wool type) was originally obtained from Owens Corning. From standard fiberglass nomenclature, the letter K represents a filament diameter of 12.7-13.96 μm and 115 stands for the spinning density in 100 yd/lbm. Thus, 115K could be interpreted as 4770 m strand per kg, type K fiberglass. During installation, the type 115K fiberglass is shredded to a light loose consistency so that it occupies twice its original volume. The filter is uniformly filled with the shredded fiberglass to slightly above the volume needed and compressed to the proper thickness and density. The overall efficiency of the PVV filter when loaded in this way is 99.92%.

Ammonium nitrate generated by the PUREX process is by far the most important compound collected by the PVV filters. Historically, ammonium nitrate has caused both the warm and hot filters to plug, thus instituting changes in processing. The 5.7F warm PVV filter plugged in January 1973 due to the presence of 5800 kg of ammonium nitrate in solution in the bottom of the filter. Filters with the current design were installed in 1974. The new filters plugged by August 1975; 600 kg of ammonium nitrate were removed, and the filters were replaced. Subsequently, process changes were instituted that decreased the amount of ammonia generated in H-Canyon and hence the amount of ammonium nitrate captured on the filters. An ammonium nitrate accumulation limit of 452 kg was placed in the Safety Analysis Report, with an operational accumulation limit of 350 kg. Exceeding this limit would require that the filters be removed and replaced. With the reduced ammonia generation rate from the process improvements, the amounts of ammonium nitrate on the filters collected have remained substantially below this limit.

Actinide accumulation on the PVV filters varies depending on whether the filter is on the warm or hot side of the canyon. The amount of uranium which accumulates is much higher in the warm PVV filter due the increased amount of uranium which is processed or stored on this side. Additionally, waste neutralization takes place in the warm side. Plutonium accumulation is greater on the hot side of the canyon because plutonium separation and storage occur there.

Actinides also accumulate on the PVV filters when process solutions are being stored. A substantial quantity of actinides is captured by the filters during periods with little to no throughput. This is understandable since the air flow being pulled through the filter entrains particles and drops of liquid from acid mists above the solutions into the filter.

The ^{235}U load curves for the hot and warm PVV filters are shown in Figure 2. The ^{235}U data is corrected for the ^{235}U which accumulates when there is no material being processed through the canyon. These load curves are combined with criticality safety limit (CSL) for $^{235}U_{equ}$ (defined in terms of the mass of ^{235}U + 1.6 times the mass of ^{239}Pu) to determine PVV filter flush intervals. The CSL for $^{235}U_{equ}$ in H-Canyon is defined as 624 g $^{235}U_{equ}$. (This the ANSI/ANS-8.1 CSL. The limit represents the critical mass of a metal sphere.) Based on the CSL, the required flush intervals have been calculated to be 23 months for the hot canyon filter and 10.7 months for the warm canyon filter. In general, these flush intervals are more limiting that intervals based on the accumulation of ammonium nitrate.

From the preceding discussion, it is apparent that the only way to extend the time between flushes was to increase the amount of material allowed on the filters by using a less conservative CSL. In the absence of defensible accumulation data, the rather conservative ANSI/ANS-8.1 CSL was chosen. At first this seems reasonable, since after flushing a uniform solution would result. However, the PVV filters are flushed with nitric acid solutions poisoned with boron. Such a solution allows kilogram amounts of $^{235}U_{equ}$ to be present without a criticality. Because the shape of the uranium mass affects the criticality limit, the distribution of fissionable material which accumulates within the PVV filters becomes an important factor.

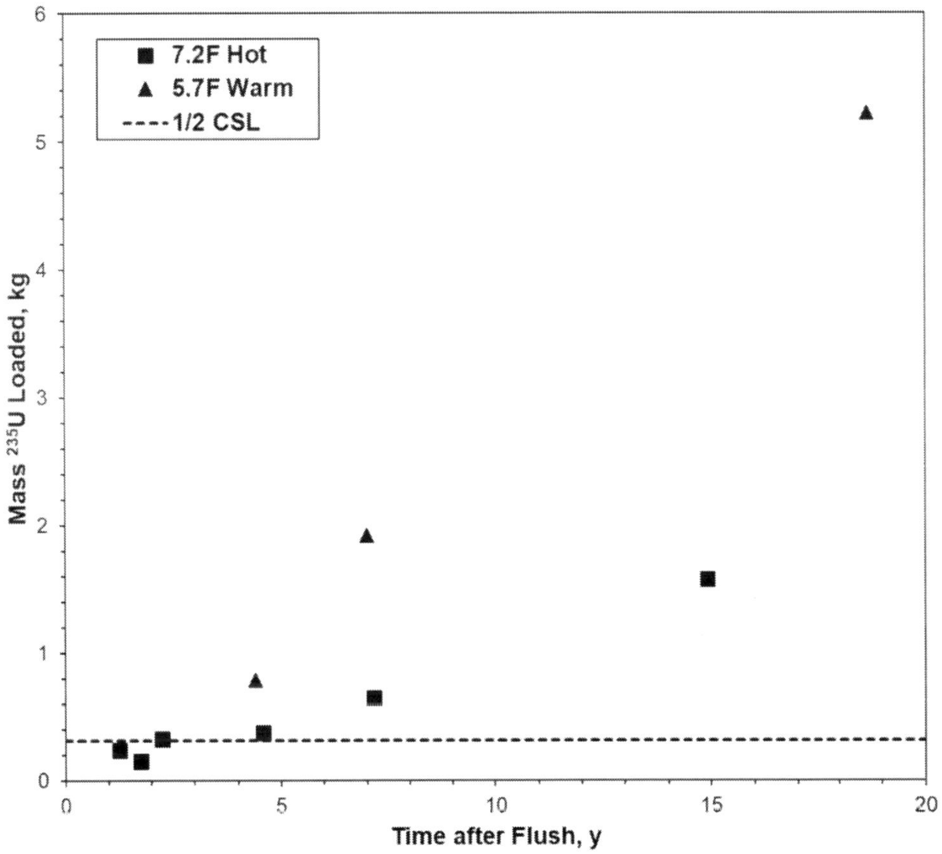

Figure 2. H-Canyon PVV filter Load Curves from ^{235}U data Corrected for the Accumulation when H-Canyon was Idle

Review of Geometric Models

Studies have shown that small particulates such as those present in the H-Canyon off-gas exhausts collect on the filter fibers as chain-like filaments called dendrites.[1,2] Most of the particulates do not penetrate the filter but instead remain on the surface of each filter layer. Thus, the uranium mass would form a disk on the upstream face of the filter, with a thickness controlled by the ratio of uranium to ammonium nitrate and the void fraction of the dendrite layer. The thickness of this disk varies as a function of the collection efficiency and the air flow rate through the filter material. If enough material accumulates in this disk the filter cakes and plugs.

Mass accumulation has been studied up to filter cake formation on successive layers of a filter medium.[3] The study showed that when filter structure was uniform with depth, clogging took place in the front portion of each layer of the filter structure. The increase in the efficiency of a fiberglass filter was studied by successively increasing its depth by adding a greater number of layers of fiberglass paper. The initial penetrations of the one to six filter layers were 41.0%, 17.5%, 7.2%, 1.73%, and 0.67%. This progression would be expected from the PVV filter as it is essentially a layered fiberglass filter. The four upstream layers of the PVV filter assembly are filled manually with loose fiberglass fibers, so its structure is uniform with depth. However, as loose fibers, neither the collection efficiency nor the air flow through these layers is uniform. The fifth and final layer is installed as a mat with a backing that subsequently is removed by heating. Therefore, the flow through and deposition on this layer should be uniform.

The general approach to analyzing filter layer nonuniformities is to assume that a portion of the filter area has a relatively sparse fiber density and that the remainder of the area is relatively dense. A fiber material balance, an air flow balance, and a pressure drop equation determine the sparse and dense surface areas, the flow rate through each of these areas, and the relative fiber density in each area. The use of a bimodal distribution bounds the effects of any arbitrary fiber density distribution on concentrating the particulate deposits.

The following analysis of filter efficiencies predicts the relative amounts of particulates that deposit on each layer of the PVV filter. The analysis also estimates the maximum amount of particulates that can concentrate in a small area on the filter surface. Deposits will concentrate in areas with relatively sparse amounts of filter fibers, due to the high air and particulate flow rates through those areas. The increase in the particulate flow through these sparse areas more than compensates for the reduction in filtration efficiency, so that there is a net increase in the deposition rate.

In a review of various filtration theories,[4] the Torgeson model[5] was found to give the best fit to data over a wide range of filter and fiber parameters. The PVV filter analysis uses the filter efficiency model developed by Torgeson with the addition of impaction introduced by Whitby to calculate deposition rates.[4,5] This model requires the input of filter layer thicknesses and densities, the filter fiber diameter, the flow rate through the filter, and the particle diameter. The model accounts for deposition of small particles by diffusion through the filter layer, and, for larger particles, interception of particles by the fibers and inertial impaction of particles on the fibers.

Torgeson Model for Particle Deposition

The Torgeson model combines a previous theory for deposition by interception and inertial impaction with a new model for deposition by diffusion and interception. (In general, small particles deposit as they diffuse into the filter material, while larger particles deposit due to interception by and impaction onto individual filter fibers.) The model treats flow through the filter as a flow past a series of single fibers. This treatment results in an exponential expression for the fraction of particles that penetrate the filter, P, as a function of the single fiber efficiency in capturing particles, η_s, and a filter solidarity factor, S.

$$P(d_p) = \exp(-\eta_s S) \tag{1}$$

The filter solidarity factor represents the number of fibers in the flow path of a particle. It is calculated from the equation

$$S = \frac{4L\rho_{f,b}}{\pi d_f \rho_{f,t}} \tag{2}$$

where L is the fiber layer height, d_f is the fiber diameter, $\rho_{f,b}$ is the average bulk fiber density, and $\rho_{f,t}$ is the theoretical fiber density.

The single fiber efficiency is the sum of fiber efficiencies for interception, η_{si}, impaction, η_p, and diffusion, η_d, with some empirical correction factors:

$$\eta_s = 0.75\eta_{si} + \eta_p + G(M)\eta_d \tag{3}$$

For interception, the single fiber efficiency is given by

$$\eta_{si} = 0.0518\frac{C_D \, Re}{2}R^{1.5} \tag{4}$$

where Re is the Reynolds number for flow past the cylinder,

$$Re = \frac{d_f \rho_{air} v_0}{\mu_{air}} \tag{5}$$

R is the ratio of the particle diameter, d_p, to the fiber diameter,

$$R = \frac{d_p}{d_f} \tag{6}$$

and C_D is the drag coefficient for flow across a single cylinder. An expression for this drag coefficient at low Reynolds numbers was originally derived by Lamb,[6] who obtained

$$C_D = \frac{8\pi}{Re(2-\ln(Re))} \tag{7}$$

The low Reynolds number assumption is not valid for all flow conditions. At higher Reynolds numbers, Schlichting presents a drag coefficient correlation based on measurements.[7] An asymptotic fit to the Schlichting data yields

$$C_D = \left[\left(10Re^{-0.77528}\right)^3 + \left(4.9122Re^{-0.16265}\right)^3\right]^{1/3} \tag{8}$$

The single fiber efficiency for particle diffusion is given by

$$\eta_d = 0.75\left(\frac{C_D Re}{2}\right)^{0.4} Pe^{-0.6} \tag{9}$$

where Pe is the Peclet number, defined by

$$Pe = \frac{d_f v_0}{D_M} \tag{10}$$

where v_0 is the superficial air velocity through the filter layer and D_M is the particle diffusivity. The particle diffusivity is calculated using Einstein's application of Stokes law, which gives[6]

$$D_M = \frac{kTC}{3\pi d_p \mu_{air}} \tag{11}$$

where k is Boltzmann's constant, T is the absolute temperature, and μ_{air} is the air viscosity.

The variable C is a velocity multiplier called the Cunningham correction factor. This factor accounts for the increase in the effective velocity due to Brownian motion in the gas. The expression for the Cunningham correction factor takes the form:[8]

$$C = 1 + 1.63\frac{2x_{air}}{d_p} \tag{12}$$

where x_{air} is the mean free path through the gas, which is approximately 0.09 μm.

Equation 3 contains a correction factor for the diffusion term, $G(M)$, which increases the single fiber efficiency slightly. The function $G(M)$ is approximated by a least-squares regression fit to graphical data.[4] The fit takes the form

$$G(M) = 1.2003 + 0.1356\log_{10}(M) + 0.0207\log_{10}^2(M) \tag{13}$$

where M is defined by

$$M = \frac{1}{2} C_D \, Re \, R^{2.5} Pe \tag{14}$$

The impaction efficiency η_p is expressed in terms of an inertial parameter, ϕ, defined by

$$\phi^{1/2} = \left(\frac{\rho_p v_0 C}{18 \mu_{air} d_f} \right)^{1/2} d_p \tag{15}$$

where ρ_p is the particle density.

From an analysis of filter data, Whitby found that the deposition efficiencies due to impaction are log normally distributed about a mean inertial parameter given by[4]

$$\phi_M^{1/2} = 1.253 Re^{-0.0685} \tag{16}$$

with the stipulation that for small Reynolds numbers

$$\phi_M^{1/2} \le 1.4 \tag{17}$$

According to the definition of log normal distributions, the impaction efficiency can be expressed as

$$\eta_p = \frac{1}{2} \mathrm{erf}\left(\left| \frac{\ln\left(\frac{\phi}{\phi_M} \right)}{\sqrt{2} \ln(\sigma_g)} \right| \right) \quad \text{for } \phi \ge \phi_M \tag{18}$$

and

$$\eta_p = 1 - \frac{1}{2} \mathrm{erf}\left(\left| \frac{\ln\left(\frac{\phi}{\phi_M} \right)}{\sqrt{2} \ln(\sigma_g)} \right| \right) \quad \text{for } \phi < \phi_M \tag{19}$$

where erf signifies the error function and σ_g is the geometric standard deviation for the impaction efficiency distribution. Whitby recommended a standard deviation of 1.65 for circular fibers.[4]

The preceding model is used to calculate the fraction of particles that penetrate each of the five layers of a PVV filter section. As stated previously, each of the first four layers is modeled as containing an area that is densely populated with fibers and an area that is sparsely populated with fibers. The overall penetration fraction is given by the following weighted average of the penetration fractions for the densely and sparsely populated areas, $P_{n,d}(d_p)$ and $P_{s,d}(d_p)$:

$$P_n(d_p) = \frac{A_d v_{0,d} P_{n,d}(d_p) + A_s v_{0,s} P_{n,s}(d_p)}{A_d v_{0,d} + A_s v_{0,s}} \tag{20}$$

where A_d and $v_{0,d}$ are the densely populated area and the superficial velocity through that area and A_s and $v_{0,s}$ are the sparsely populated area and the superficial velocity through that area. The partition of areas between densely populated and sparsely populated regions is treated as a variable input parameter. The fifth and final filter layer is modeled as a uniform filter, as explained previously.

Deposition Distribution Calculation

The cumulative distribution for the deposited particles is obtained by summing contributions from particles of different diameters. The distribution of particles of a particular diameter d_p on the filter layers can be calculated directly from the results of the Torgeson model. The fraction that deposit on the nth filter layer, $X_n(d_p)$, is just the fraction of particles that penetrate the upstream, or (n-1)st layer, but do not penetrate the nth layer. In other words,

$$X_n(d_p) = \left[1 - \sum_{m=1}^{m=n-1} X_m(d_p)\right]\left[1 - P_n(d_p)\right] \quad (21)$$

where $P_n(d_p)$ is the fraction of particles of diameter d_p that penetrate the nth layer.

The total fraction deposited on the nth filter layer, $\overline{X}_n$, is calculated by integrating the product of the deposited fractions for any given particle diameter and the probability density function for the particle diameters. To maintain a material balance, the probability density function is defined in terms of mass fractions rather than numbers of particles.

In addition to the overall particle distribution on each filter, the distribution of particles between the densely and sparsely populated portions of the first filter layer is of interest. Particles will tend to congregate on the first filter layer where the flow rate through this filter is the highest. Consequently, the size of the sparse region of this filter layer and the fraction of particles that deposit in this region determine the maximum uranium concentration anywhere in the PVV filter assembly. The fractions of particulates that deposit on the dense and sparse regions of the first filter layer, $X_{1,d}(d_p)$ and $X_{1,s}(d_p)$, are given by

$$X_{1,d}(d_p) = \frac{A_d v_{0,d}}{A_d v_{0,d} + A_s v_{0,s}}\left[1 - P_{1,d}(d_p)\right] \quad (22)$$

and

$$X_{1,s}(d_p) = \frac{A_s v_{0,s}}{A_d v_{0,d} + A_s v_{0,s}}\left[1 - P_{1,s}(d_p)\right] \quad (23)$$

These quantities are integrated to obtain the overall deposition fraction in these regions in the same manner used to obtain the deposition fraction on each filter layer.

Particle Size Distribution Calculation

The calculation of overall deposition fractions requires an assumption about the form of the probability density function for the particle size. Airborne particulates generated by nuclear processing typically exhibit a log normal distribution of particle diameters when the distribution is weighted by mass.[9] The recommended particle size distribution for fuel handling facilities is 4.0 μm, with a geometric standard deviation of 1.8.[9] (The geometric standard deviation is defined as the ratio of the mean diameter to the diameter one standard deviation from the mean. Use of a geometric standard deviation implies that the diameters are log normally distributed.) The recommended value is an average of 61 different measurements which include processing environments like the canyons. Fabrication operations produce particles with mean diameters between 3 and 5 μm, with a single low value of 0.34 μm for chemical recovery facilities.[10] Not surprisingly with the large air flow, predicted collection efficiencies using the Torgeson model with a 0.34 μm particle diameter failed to approach the observed collection efficiency of 99.92%.[1] Consequently, the recommended mean particle size of 4.0 μm with a geometric standard deviation of 1.8 was used in the analysis.

The particles captured by the PVV filters are assumed to have the density of a saturated ammonium nitrate solution or 1725 kg/m^3. The assumption was made because ammonium nitrate is the main species generated by the H-Canyon process and captured by the filters. Since ammonium nitrate is very hygroscopic, the droplets and particles formed over the process solutions would become saturated. The assumed density encompasses the density of saturated uranium nitrate solutions of 1669 kg/m^3. The loss of ammonium nitrate due to decomposition and evaporation after capture by the filter has been neglected in this model.

Calculation of Dense and Sparse Areas, Densities, and Velocities

To complete the analysis, a calculation of the relative areas, bulk densities, and velocities for flow through the densely and sparsely populated filter regions is needed. The calculation is based on a mass balance for the filter material, an air flow balance, and a pressure balance. The filter material balance is

$$A_d \rho_{f,b,d} + A_s \rho_{f,b,s} = (A_d + A_s)\rho_{f,b} \quad (24)$$

where $\rho_{f,b}$, $\rho_{f,b,d}$, and $\rho_{f,b,s}$ are the average bulk density in the filter layer, the bulk density in the dense region, and the bulk density in the sparse region.

The air flow balance takes the form

$$A_d v_{0,d} + A_s v_{0,s} = (A_d + A_s) v_0 \quad (25)$$

Finally, the pressures in the two regions must be equal:

$$\Delta P_d = \Delta P_s \quad (26)$$

The pressure drop is calculated as the product of a modified formula for flow past a single fiber and the number of fibers in the flow path[11]

$$\Delta P = \frac{1}{2} C_D f_f \rho_{air} v_0^2 S \quad (27)$$

The parameter f_f is a factor that accounts for interference from other fibers. It can be approximated by[4]

$$f_f = 1 + 30\frac{\rho_{f,b}}{\rho_{f,t}} \quad (28)$$

Calculation of Effect of Particulate Loading on Deposition and Pressure Drop

The preceding analysis applies to a filter that is not burdened by particulate deposits. To examine the effect of particulate loading on deposition, it is necessary to assign a density and an effective fiber diameter to the deposited particles. Previous studies have modeled the deposited particles as fibrous dendrites, with a density and effective diameter governed by the flow field around the filter fibers.[1,2] This study takes a simpler approach. The effective diameter for the deposits is set equal to the fiber diameter. The bulk fiber density is calculated by multiplying the theoretical fiber density by the total combined volume of the deposited particles and the filter fibers and dividing by the filter layer volume. The density calculation assumes that there is no volume increase due to particle deposition. Separate densities are computed for the sparse and dense regions of each filter layer.

The particle deposition model distributes the particles between the sparse and dense areas of the filter layers in proportion to the collection efficiencies of and flow rates through these two areas. If the sparse and dense areas form single distinct regions, particle deposition would increase the density of each region without changing its area. However, it is more realistic to assume that the sparse regions are widely dispersed throughout the cross-sectional flow area, so that these regions are in intimate contact with regions with higher fiber density. Particles that travel through small, dispersed regions of sparse density are more likely to deposit on fibers in adjacent high-density regions than on fibers in the sparse-density region. Such deposits would decrease the area of the sparse region rather than increase its density.

A detailed model that partitions deposits between increases in the density and decreases in the area of sparse fiber regions is beyond the scope of this study. In lieu of such a model, the deposition calculation assumes that, for each filter layer, an initially undetermined fraction of the particles that deposit in the sparse region increases its density and that the remainder of the depositing particles reduces its flow area. To solve for this fraction, an iteration is performed to equate the pressure drops across the sparse and dense regions, while keeping constant the ratio of the velocity through the sparse region to the average velocity. This assigns most of the deposition in the sparse region to a decrease in its flow area, but allows a sufficient increase in the sparse region density to prevent a physically implausible acceleration of the flow as the filter accumulates deposits.

Particulate loading affects not only the deposition distribution but also the flow rate through the filter. As particles load onto the filter layers, the pressure drop across the filter increases and the flow rate decreases. The total pressure drop across the PVV filter, ΔP_{PVV}, and the dampers and the remaining vent system piping, ΔP_{sys}, equals the negative gauge pressure just upstream from the fan, ΔP_{fan}:

$$\Delta P_{fan} = \Delta P_{sys} + \Delta P_{PVV} \tag{29}$$

At the design flow rate of 1.18 m^3/s through the PVV filters, or 0.59 m^3/s through either the upper or lower half of a filter assembly, the fan generates a vacuum of 4.73 kPa. Of this total, the filter pressure drop for a clean filter accounts for 0.75 kPa, according to data obtained from the 5.7F warm canyon filter, and the initial pressure drop across the system piping takes the remaining 3.98 kPa. As the filter accumulates particles, the pressure drop across the filter increases and, consequently, the flow rate decreases. Assuming that the fan is a straight-blade type, the fan still pulls the same 4.73 kPa at the reduced flow rate.[12] Because the system piping consists primarily of fittings, bends, and tees which have velocity head losses, the pressure drop in this piping decreases approximately as the square of the velocity. The pressure drop across the filter is proportional to the flow rate to a power n, which can be determined from the pressure drop equation (Equation 29) and the drag coefficient correlation (Equation 8). Due to the small fiber diameters, the Reynolds numbers for the filter flow are in the low asymptote range where the drag coefficient is proportional to the velocity to the – 0.77528 power. Thus, for the filter flow, the pressure drop is proportional to the velocity to the 2 - 0.77528 or 1.22472 power.

To write the equation for the pressure drop across a loaded filter, it is convenient to define a variable y_Q equal to the ratio of the flow rate for the loaded filter to the flow rate for a clean filter. In terms of y_Q the pressure drop equation becomes

$$\Delta P_{fan} = \Delta P_{sys,i} y_Q^2 + f_{load} \Delta P_{PVV,i} y_Q^n \tag{30}$$

where $\Delta P_{PVV,i}$ is the pressure drop across a clean filter, $\Delta P_{sys,i}$ is the pressure drop across the remaining system piping when the filter is clean, f_{load} is the ratio of the pressure drop across a loaded filter to the pressure drop across a clean filter at the same flow rate,

$$y_Q = \frac{Q}{Q_i} \tag{31}$$

and

$$n = 2 - 0.77528 = 1.22472 \tag{32}$$

The preceding equations yield the following result for the flow rate as a function of particle loading

$$Q = Q_i \sqrt{\frac{\Delta P_{fan} - \Delta P_{PVV}}{\Delta P_{sys,i}}} \tag{33}$$

where ΔP_{PVV} is evaluated using the particle loading model.

As the filter becomes loaded with particles, the filter pressure drop increases and the flow rate through the filter decreases until, eventually, the fan is unable to pull air. This condition can be analyzed using a perturbation analysis that considers the effects of small changes from equilibrium flow. If the filter were absent, a momentary increase in the flow rate would raise the pressure drop in the system piping and lower the pressure (or increase the gauge pressure) just upstream from the fan. The gauge pressure pulled by a fan typically drops slightly as the flow rate increases,[15] so the fan would compensate by taking in less air. The net result is that the system is stable with respect to small flow perturbations.

With a filter present, this situation can change. Recall that the fan gauge pressure remains nearly constant as the flow rate changes, so an increase in the system pressure drop decreases the pressure drop across the filter, and vice versa. Thus, if the change in the filter pressure drop exceeds the change in the system pressure drop, a momentary acceleration of the fan would decrease, not increase, the pressure drop in the system piping. The upstream response would be to take in an increased flow of air. However, due to frictional resistance, the filter flow rate would not adjust nearly as rapidly to flow perturbations as either the fan or the piping. As a result, the system pressure would decrease still further, and a flow instability would ensue. The condition of marginal stability can be calculated by equating the derivatives of the system and filter pressure drops with respect to the flow rate. This gives

$$\left.\frac{d\Delta P_{sys,i} y_Q^2}{dy_Q}\right|_{y_Q = y_{Q,min}} - \left.\frac{df_{load}\Delta P_{PVV,i} y_Q^n}{dy_Q}\right|_{y_Q = y_{Q,min}} = 0 \tag{34}$$

where $y_{Q,min}$ is the minimum flow rate at marginal stability.

Solution of this equation yields

$$f_{load} = \frac{2\Delta P_{sys,i}}{n\Delta P_{PVV,i}} y_{Q,min}^{2-n} \tag{35}$$

Finally, substitution of this expression for f_{load} in Equation 30 gives, for the minimum flow rate

$$y_{Q,min} = \sqrt{\frac{n\Delta P_{fan}}{(n+2)\Delta P_{sys,i}}} \tag{36}$$

For the PVV filters, using a clean filter pressure drop of 3 inches of water, the minimum flow rate for a loaded filter from this analysis is 67.2% of the initial flow rate through a clean filter, or 0.396 m^3/s through one-half of a filter assembly. The corresponding filter pressure drop for the loaded filter, from Equation 33, is 2.94 kPa.

Calculation of Particulate Deposition Distribution

A computer code was written to calculate the distribution of deposited particles. This code applies the Torgeson particle deposition model, a log normal particle size distribution, and a pressure drop model to calculate the distribution of particles on the filter layers. The code first calculates the initial distribution of particle deposits on a clean filter assembly. This calculation includes three iterations, an inner iteration to equate the pressure drops for the dense and sparse regions of each filter layer, a second iteration to match the calculated filter efficiency to the reported efficiency of 99.92%,[9] and an outer iteration to set the calculated pressure drop across all five filter layers at the measured value of 0.75 kPa. Variables for these iterations are the ratios of the velocities in the sparse and dense regions to the average velocity in each filter layer and the ratio of the dense region bulk density to the sparse region bulk density. Two different density ratios are used, one for the first four filter layers and another for the final filter layer.

The second part of the calculation analyzes the effect of particle loading on deposition distributions, pressure drops and flow rates. The loading calculation involves two iterations. An inner iteration equates pressure drops for the dense and sparse regions of each filter layer. Variables for this iteration are the ratio of the dense region velocity to the average velocity and the fraction of the deposits in the sparse region that reduce its area rather than increase its density. An outer iteration updates the flow velocity to match the calculated pressure drop for each incremental increase in the weight of particles that are loaded.

To achieve numerical stability, most of the iterations are damped; this necessitates the use of many iterations to obtain accurate results. An exception is the iteration to update the flow velocity. This iteration was purposely left undamped to simulate physical interactions between pressure and flow rate. For every increase in the assigned weight of loaded particles, the flow rate and pressure are alternately updated five times to check for stability. Instabilities were detected at particle loads ranging from just over 272 kg to over 318 kg for half of a filter assembly, with the lower limit calculated for small increments in the load weight and the higher limit occurring for larger increments. Usually, the instabilities started with pressure drop and velocity oscillations that grew as the load increased. However, at load weight increments of 0.0113 and 0.0226 kg, both the pressure drop and the velocity decreased smoothly and monotonically, ending at loads of 281 kg and 283 kg, respectively. These results may indicate that these calculation step sizes represent a type of harmonic frequency at which the calculation converges on the true value of the instability. Additional discussion of pressure drop and flow rate results appears in the following section.

Based on observations of pressure drop instabilities, a step size of 0.0226 kg was selected for the analysis. Up to the point of instability, variations of particle deposition distributions, pressure drops, and flow rates were not sensitive to the step size.

Results of the Particle Deposition Calculations

The two results that pertain to the deposition of $^{235}U_{equ}$ on the filters are the distribution of particulates among the five filter layers of a section and the fraction of particulates that can accumulate in a small area of a single filter layer. The first result controls the mass of uranium that can form as a flat disk with a surface area equal to the flow area of the filter. The second result limits the mass of uranium that accumulates in a small area.

Figure 3 depicts the variation of the cumulative fractions of the particles that deposit on each filter layer as a function of the weight of particles loaded. As the results in this figure show, most of the particles deposit on the first, upstream filter layer, and nearly all the deposition occurs on the first two layers. The particle distribution changes only slightly from its initial value and in fact appears to asymptotically approach a limit as the loading increases.

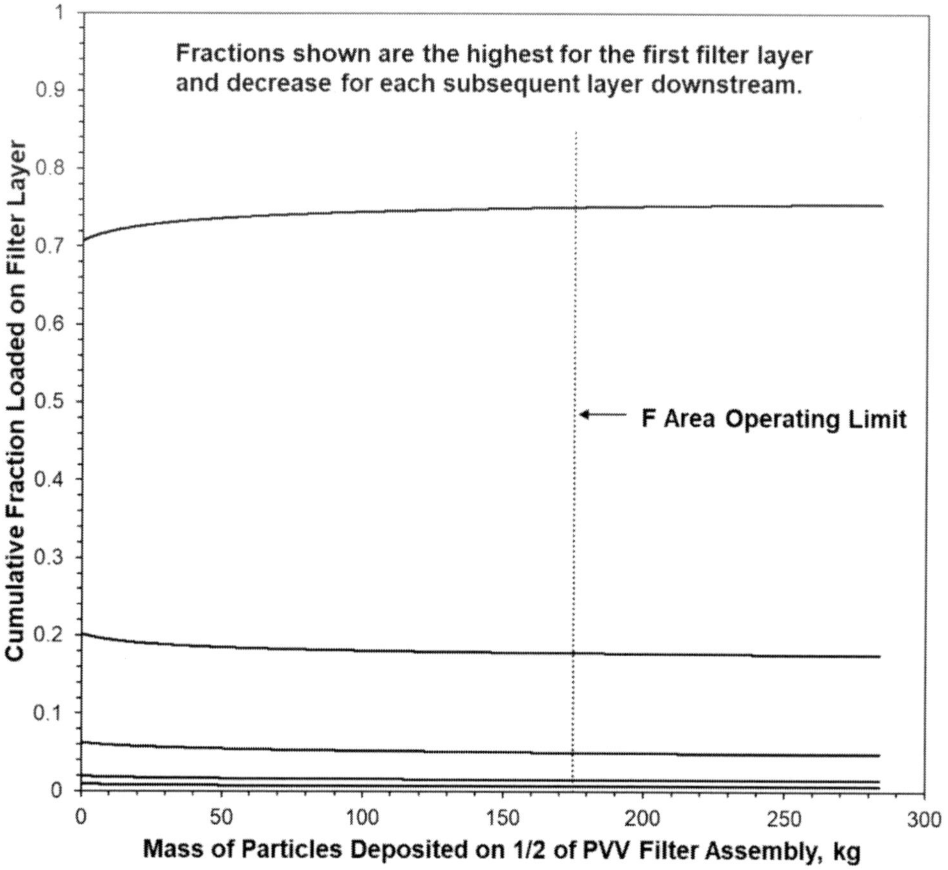

Figure 3. Cumulative Loading of Five Filter Layers within a Section of the PVV Filter

By far the greatest concentration of deposited particles is in the sparse fiber region of the first filter. As a limiting case, it can be assumed that if a critical mass of fissile material accumulates, that mass will accumulate in particles at this location. The particle concentration in other regions will be too dispersed for the actinide accumulation to be treated as a moderated metal sphere as is done for the ANSI/ANS criticality limit.

The fraction of particles that deposit in the sparse region decreases with total particle loading, as illustrated by the loading transient depicted in Figure 4. A bounding maximum for the fraction of particles in the sparse was determined by calculating the initial fraction of particles that deposit in the sparse region of the first filter as a function of the fiber density distribution in that filter, as specified by the densely and sparsely populated filter areas. As Figure 5 shows, no more than 4% of the particles deposit in the sparsely populated region of the first filter.

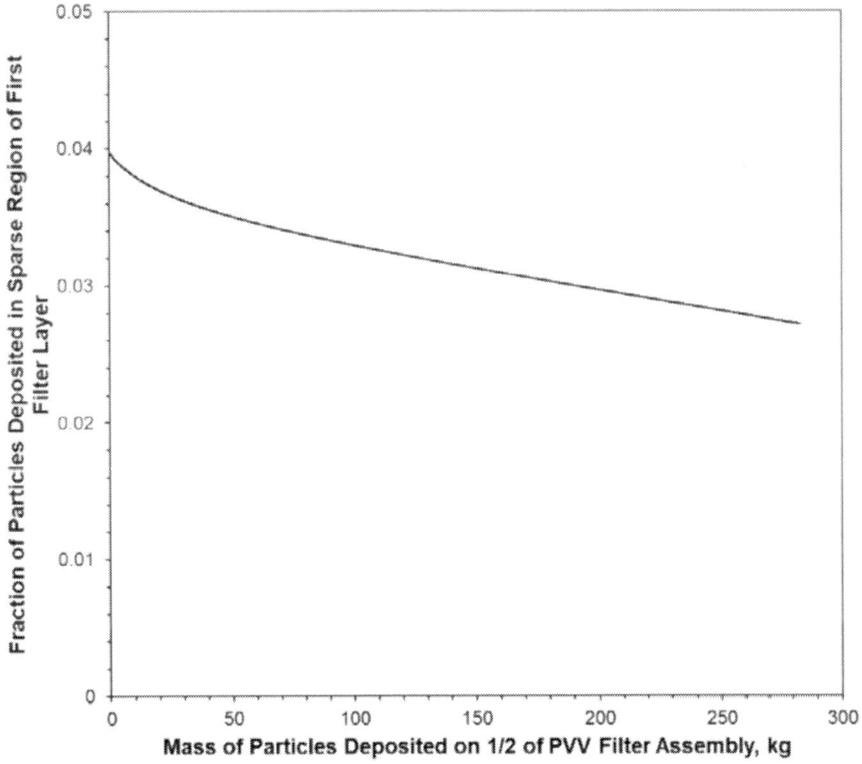

Figure 4. Typical Variation of Fraction of Particles Deposited in the Sparse Region of the First Filter Layer with Cumulative Deposition

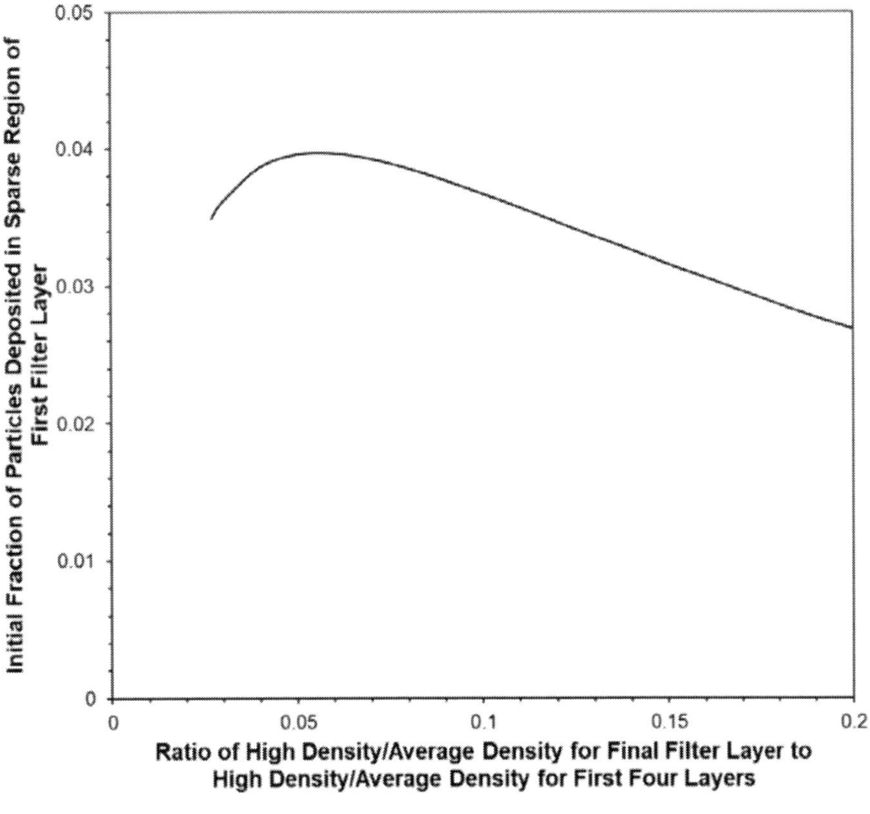

Figure 5. Initial Fraction of Particles Deposited in the Sparse Region of the First Filter Layer as a Function of Filter Density Distribution

In addition to loading distributions, the particle deposition model calculates pressure drops and air flow rates through the filter as functions of the load weight. Figure 6 shows the variation of the flow rate and depicts changes in the filter pressure drop with loading. The flow rate decreases from its initial value of 0.59 m^3/s to 0.39 m^3/s at the particle load limit, while the pressure drop increases from 0.75 kPa to 2.96 kPa. These calculated limits agree closely with the results derived from the stability analysis, which were a limiting flow rate of 0.41 m^3/s and a limiting pressure drop of 2.94 kPa. The flow rate and the pressure drop vary approximately linearly with the weight of particles loaded up to the limit, at which point both drop precipitously, indicating an abrupt transition at the load limit. The decrease in the pressure drop is an artifice of the calculation method, which requires the pressure drop across the filter to be in equilibrium with the flow rate, even past the maximum stable flow limit.

Comparisons between measured and calculated pressure drops and flows demonstrate that the deposition calculation accurately models flow within the filter. Measurements just prior to and after the November 1998, filter flush in the 5.7F warm canyon recorded that flushing resulted in a pressure drop change of 0.22 kPa, from 0.97 to 0.75 kPa. The amount of ^{235}U removed from the filter was 0.209 kg. The amount of ammonium nitrate for this flush was 50.6 kg. This weight corresponds to a calculated pressure drop of 1.10 kPa. The change from the initial pressure drop for the clean filter is 0.35 kPa, which is 58% higher than the measured change of 0.22 kPa. In other words, the calculated pressure drop is 23% higher than the 0.97 kPa measured prior to when the filter was flushed.

A second comparison can be made between the measured plugging limit for the PVV filters and the calculated load limit for stable operation of the fan. The measured filter plugging limit is 600 kg for the entire filter assembly, which translates to 300 kg for one-half of the assembly. The calculated load limit is 283 kg, or 6% lower. While this comparison is indirect in that it presumes that the plugging limit is equivalent to the stability load limit, it offers additional support that the flow and pressure drop models in the deposition model are accurate.

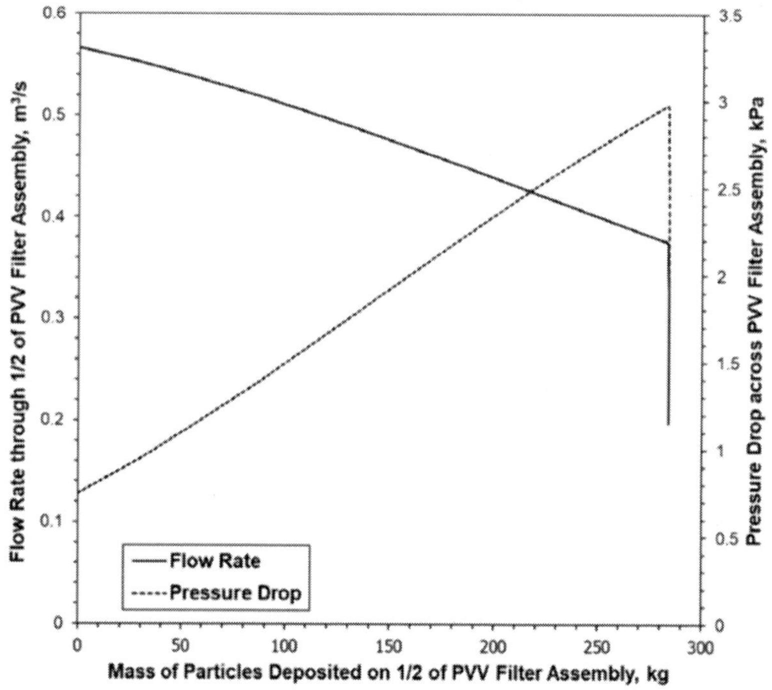

Figure 6. Variation of Air Flow Rate through and Pressure Drop across the PVV Filter with Particle Deposition

Conclusions

A geometric calculation was used to predict the spatial distribution of particles depositing on the H-Canyon PVV filters. The calculation was based on the Torgeson model for particle deposition on filters.[5] The calculation predicted that most of the particles deposit on the first of five filter layers, with the highest concentration of particles located in the portion of the first filter layer that was assumed to be sparsely populated with fibers. It was determined that no more than 4% of the total particle matter would deposit in this highly concentrated region.

The CSL for the PVV filters was changed based on this result. The previous CSL was set at 624 g $^{235}U_{equ}$, equal to the critical mass of a moderated metal sphere. For the revised CSL, the critical mass was based on only the fraction of the actinide that could accumulate in one concentrated region. The revised CSL was increased to 15 kg $^{235}U_{equ}$, in principle calculated by dividing the old CSL, rounded down to 600 g, by the highly concentrated deposition fraction of 0.04. At the increased CSL, the determining factor for filter flushing becomes the accumulation of ammonium nitrate and filter pluggage rather than criticality.

A filter pluggage calculation based on the Torgeson model gave a maximum particle loading of 283 kg of ammonium nitrate on each of the two PVV filters. This is in close agreement with the observed pluggage limit of 300 kg.

References

1. Payatakes, A. C. (1977) "Model of Transient Aerosol Particle Deposition in Fibrous Media with Dendritic Pattern," AIChE J., 23(2), 192-202.
2. Payatakes, A. C., and L. Gradoń (1980) "Dendritic Deposition of Aerosol Particles in Fibrous Media by Inertial Impaction and Interception," Chem. Eng. Sci., 35(5), 1083-1096.
3. Japuntich, D. A., J. I. T. Stenhouse, and B. Y. H. Liu (1997) "Effective Pore Diameter and Monodisperse Particle Clogging of Fibrous Filters", J. Aerosol Sci., 28(1), 147-158.
4. Whitby, K. T. (1965) "Calculation of the Clean Fractional Efficiency of Low Media Density Filters," ASHRAE J., September, 56-65.
5. Torgeson, W. L. (1964) "The Theoretical Collection Efficiency of Fibrous Filters Due to the Combined Effects of Inertia, Diffusion and Interception," Paper No. J 1057, Applied Science Div., Litton Systems, Inc., St. Paul, Minnesota.
6. Chen, C. Y. (1955) "Filtration of Aerosols by Fibrous Media," Chem. Rev., 55(3), 595-623.
7. Schlichting, H. (1968) Boundary-Layer Theory, McGraw-Hill, New York (1968), p. 17.
8. Cunningham, E. (1910) "On the Velocity of Steady Fall of Spherical Particles through Fluid Medium," P. Roy. Soc. A-Math. Phys., 83, 357-365.
9. Dorrian, M. D., and M. R. Bailey (1996) "Particle Size Distributions of Radioactive Aerosols Measured in the Workplace and the Environment," J. Aerosol Sci., 27(Supp. 1), S439-S440.
10. Dorrian, M. D., and M. R. Bailey (1995) "Particle Size Distributions of Radioactive Aerosols Measured in Workplaces," Radiat. Prot. Dosim., 60(2), 119-133.
11. Perry, R. H., and D. W. Green, eds. (1984) Perry's Chemical Engineers' Handbook, 6th ed., McGraw-Hill, New York, p. 6-23.

577507 Solvent Extraction Flowsheet-Level Modeling and Simulation with Radiolysis

Solvent Extraction Flowsheet-Level Modeling and Simulation with Radiolysis

Valmor F. de Almeida †, Kevin L. Lyon‡, and Taha Azzaoui *

† *University of Massachusetts Lowell, Dept. of Chemical Engineering, Nuclear Program, Lowell, MA, 01854, valmor_dealmeida@uml.edu*

‡*Idaho National Laboratory, Aqueous Separations and Radiochemistry, Idaho Falls, ID, 83415, Kevin.Lyon@inl.gov*

* *University of Massachusetts Lowell, Dept. of Computer Science and Mathematical Sciences, Lowell, MA, 01854, taha_azzaoui@student.uml.edu*

ABSTRACT

Hydrometallurgical separation processes have played a critical role in the nuclear fuel cycle since the inception of nuclear energy. Solvent extraction has been utilized at the industrial scale for the recovery of uranium from ore, uranium and plutonium recovery from irradiated nuclear fuel, and nuclear waste management. Despite widespread application in the nuclear fuel cycle, development of robust models that aid in the design, operation, and troubleshooting of advanced separation processes remains one of the most significant challenges in separation science. The need for advanced modeling and simulation capabilities is a critical component to enable successful deployment of advanced nuclear fuel cycle technologies. Furthermore, developing advanced separation technologies is a crucial step towards demonstrating proof-of-principle concepts for innovative actinide chemistry. As such, the present work is aimed at developing an extensive solvent extraction flowsheet simulation capability using chemical non-equilibrium modules employed in a variety of cross-cutting applications including nuclear separations, safeguards, materials protection, special materials for national security and non-proliferation, critical materials, mining, and metallurgy applications.

A modeling framework is currently under development that couples multiple stages of solvent extraction equipment in a counter-current cascade configuration to solve time dependent non-equilibrium mass balances. A rigorous chemistry module has been implemented using available literature data for n-tributyl phosphate/dodecane and nitric acid solutions of actinides and fission products. Actinide and fission product concentrations are computed from fuel depletion codes such as ORIGEN. The module includes radiolysis reactions for H_2O and HNO_3; actinide oxidation-reduction reactions; complexation reactions for actinides, H_2O and HNO_3; and n-tributyl phosphate degradation reactions by radiolysis and hydrolysis. Overall, the module considers 152 reactions for a total of 98 chemical species. The resulting, time-dependent, highly nonlinear system of ordinary differential equations are solved using the computational framework Cortix, a python library for system-level modeling.

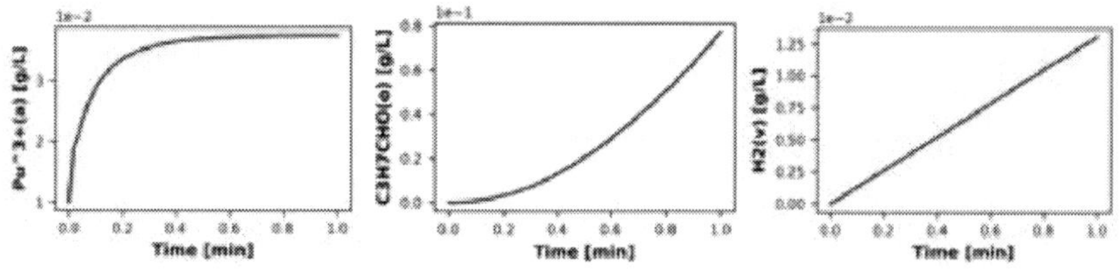

578159 Model of Helium Bubble Density and Growth in Tungsten from Equations of State

Helium bubbles in plasma-facing tungsten are responsible for degradation of material properties and deformation of the surface at the length scale of 10–100 nm. Simulations have shown that helium bubbles that are relatively far from the tungsten surface consistently grow in a well-defined band of helium/vacancy ratios. The *mean* bubble density for a given bubble size is also well-defined. We previously suggested an empirical correlation for bubble density, namely $n_{He} = 5\, n_V^{0.86}$, where n_{He} is the number of helium atoms in the bubble and n_V is the number of vacancies. Here, we explore the possibility of a theoretically-based correlation for bubble density based on material properties and plasma-related defect characteristics.

562684 On the Onset of 'Fuzz' Formation in Plasma-Facing Materials. I. Surface Morphological Evolution Model

Tungsten (W) and tungsten alloys are widely viewed as the most promising plasma-facing material (PFM) candidates for divertor and first-wall systems in a nuclear fusion reactor. However, experiments aiming to investigate PFM performance under continuous operation and high first-wall temperature have shown that a nanostructure with a fuzz-like morphology develops on the W surface under the operating conditions of temperature, helium (He) impact energy, and He flux expected for ITER's divertor. Formation of such fuzz nanostructure may adversely influence the reactor performance and operation.

Here, we focus on obtaining a fundamental understanding of the initial stage of fuzz formation by modeling the surface morphological evolution of helium-ion-irradiated tungsten considered as a PFM based on an atomistically-informed, continuous-domain predictive model that we have developed. The constitutive equations of the model are parameterized based on analysis of large-scale molecular-dynamics (MD) simulation results; for the He-implanted W layer, the stress level is calculated using heterogeneous elastic inclusion theory and the state of stress in this region is taken to be biaxial compression consistent with MD simulations of the deformation state of a He-implanted W region. According to this model, we have conducted self-consistent numerical simulations of the dynamics of the He-implanted W surface morphology and compared the simulation results with experimental measurements. Under the experimental conditions of exposure of ITER-grade W at 1113 K to a medium-flux RF plasma source (~ 3×10^{20} $m^{-2}s^{-1}$) of 75 eV He, the predicted surface morphology shows a good qualitative and quantitative agreement with the experimental data for a He plasma exposure time of 80 min. The strong quantitative agreement of the mean and width of the height distribution functions for the experimental and computed W surface morphologies establishes that our model can predict successfully the initial growth rate of nanotendrils emanating from the surface, which is the precursor to fuzz formation. In addition, we have conducted a linear stability analysis to examine the morphological stability of the planar surface of the helium-implanted tungsten and the simulation results are consistent with the predictions of the linear stability theory. We also present a parametric sensitivity analysis for the effects on our model predictions for the average nanotendril spacing and the nanotendril growth rate of three key model parameters, namely, the He concentration in the He-implanted tungsten, the He nanobubble size, and the elastic moduli of the damaged W in the nanobubble region. Furthermore, we have investigated the dependence of the nanotendril width and arrangement on the surface temperature for He-ion energy below (75 eV) and above (250 eV) the sputtering threshold, and compared the simulation results with measurements from carefully designed experiments.

568488 Helium Aggregation and Surface Morphology Near Surfaces and Grain Boundaries in Plasma-Facing Tungsten

Helium's effective insolubility in metals has led to a host of problems with helium embrittlement and bubble formation. In fusion reactors, most bubbles form near the surface because of the high flux of ions there, yielding significant surface deformation. Efforts to understand the mechanisms relevant to helium-induced surface evolution are many, but simulations of these phenomena are often hampered by finite size effects and unrealistic flux or fluence values. We discuss here the effects of helium flux, surface orientation, and grain boundary orientation on helium retention and transport as well as tungsten surface evolution in situations intended to be representative of the ITER divertor under pure helium plasma. In particular, we focus on the effects of grain boundaries that intersect the surface, observing that helium collects at the grain boundaries because of reduced rates of transport of helium clusters on grain boundaries compared to the tungsten bulk. The orientation of the surface near the grain boundary is an important aspect of the predicted surface evolution, as it determines whether dislocations emitted by bubbles trapped at the interface will escape to the surface directly above the grain boundary or whether they will be emitted and form surface features farther away.

562703 On the Onset of 'Fuzz' Formation in Plasma-Facing Materials. II. Effects of Subsurface Bubble Dynamics, Temperature, and Elastic Moduli of Damaged Tungsten

Fuzz formation in helium-implanted tungsten (W) is a complex, multi-physics phenomenon resulting from the combined effects of driven surface diffusion, subsurface bubble dynamics, bubble bursting, loop punching, anisotropies of material properties, changes in material thermophysical properties in the damaged tungsten as a function of helium (He) content, and many more. Here, we focus on understanding the individual effects on the He-implanted W surface morphology of three such important physical phenomena and physical dependences, namely, subsurface bubble dynamics, temperature dependence of material properties, and changes in material properties due to helium implantation.

We report a systematic study of the subsurface He bubble dynamics based on a continuum-domain, spatially-distributed drift-diffusion-reaction cluster-dynamics model of He cluster and bubble evolution, as well as the retention of helium in low-energy He plasma-exposed W. Our cluster-dynamics simulator, Xolotl, has been benchmarked based on large-scale molecular-dynamics (MD) simulation results and experimental measurements available in the literature. For realistic predictions of the sub-surface He concentration, which reaches a steady state at an early stage of He-plasma exposure, a simplified helium bubble bursting model is incorporated into Xolotl to take into account the gas release occurring when a bubble is near the tungsten surface. A systematic comparison of the modeling predictions with experimental measurements will be presented and strategies for further improvement of the bubble bursting model implemented in Xolotl will be discussed.

We will also present a comparison of the predictions of our experimentally validated, continuum-scale model of surface morphological evolution for He plasma-exposed tungsten with experimental measurements under different temperature conditions. For this purpose, a set of ITER-grade W specimens were mechanically polished, stress relieved at 1000 °C, and then exposed to RF-generated He plasmas (incident ion energy of 75 eV and flux of 8.5×10^{20} $m^{-2}s^{-1}$) at the same ion fluence (1.53×10^{25} m^{-2}) but at temperatures varied between 472 °C and 786 °C. Following the plasma exposure, the W surfaces are characterized and compared with our computed morphologies under identical plasma exposure conditions.

Furthermore, we report the effects on the fuzz formation of the changes in the elastic moduli of the He-implanted W with respect to those of the undamaged tungsten matrix. The elastic moduli are determined through systematic MD simulations of straining tests on single crystalline tungsten containing nanopores filled with the proper amount of He to form helium nanobubbles, which are used as models of plasma-exposed tungsten in order to assess the effect of He-ion irradiation on tungsten's elastic properties. These atomistically computed elastic moduli are incorporated into our surface evolution model to assess effects of the varying elastic properties of plasma-exposed tungsten on its surface morphology. A roadmap for the study of the combined effects of the three aforementioned physical phenomena and dependences on the predictions of our surface evolution model and a discussion on the incorporation into the model of other important physics that affects the fuzz formation process also will be presented.

562758 Mechanisms of Rubidium and Krypton Release from Zeolite Getters

Preventing the release of radioactive krypton (^{85}Kr) to the atmosphere is a major obstacle in nuclear fuel reprocessing. One of the methods used to store radiokrypton is to use zeolites with small "windows" between pores, such as sodalite (framework code **SOD**), zeolite 5A (**LTA**), and zeolite 13X (**FAU**). Systems based on this concept load krypton into the zeolite at high temperature and pressure, which is then rapidly cooled. The resulting decrease in the window size encapsulates the gas, in principle preventing its release. However, the decay product of ^{85}Kr, ^{85}Rb, is highly corrosive and can damage the metal containers in which the zeolites are stored. For example, researchers at Oak Ridge National Laboratory recently investigated the content of five canisters of zeolites that were used years ago to encapsulate radiokrypton, finding two of these canisters cracked and the insides severely corroded. Such severe corrosion suggests the presence of rubidium. This raises a question: did the krypton decay while still encapsulated followed by diffusion of the rubidium out of the sodalite cages, or did krypton leave the sodalite cage and subsequently decay to rubidium? We answer this question by studying the rate at which krypton atoms, rubidium atoms, and rubidium ions traverse the six-membered rings of sodalite cages by way of density functional theory calculations and the nudged elastic band method combined with classical transition state theory. We also investigate how the presence of non-sodium charge-compensating cations or water affects the rates of diffusion.

562774 Defect Interaction Energetics and Stress State of Plasma-Exposed Tungsten

Tungsten (W) is a promising plasma-facing component (PFC) material in nuclear fusion reactors because of its excellent thermomechanical properties and low sputtering yield. In tungsten, interstitial He atoms are very mobile and aggregate to form clusters of various sizes. Small, mobile helium clusters (He_n, $1 = n = 7$) are attracted to the tungsten surface due to an elastic interaction force that drives surface segregation, and their diffusional transport mediates the dynamics of surface morphology and near-surface microstructure.

Here, we report results for helium-related defect interaction energetics near surfaces of tungsten that has been exposed to different levels of He ion irradiation based on atomistic simulations according to a reliable many-body interatomic potential. At higher helium fluence, mobile helium clusters are subjected to cluster-defect interactions in addition to cluster-surface interactions, which complicate cluster dynamics beyond the dilute limit of helium content in the PFC material. Based on configurations generated by large-scale molecular-dynamics (MD) simulations of implanted helium evolution in plasma-exposed tungsten, we examine the effects of varying helium fluence due to increased plasma exposure byconducting systematic molecular-statics (MS) computations of small helium cluster energeticsnear the W surface as a function of distance (depth) of the cluster center from the surface on a grid of lateral locations on the surface. We analyze the defect interactions that mediate the energetics of small helium clusters migrating to the surface, taking into account that the migrating cluster also is subjected to the stress fields generated by larger helium bubbles, as well as other small clusters. The outcome of this analysis is the systematic parameterization of mobile helium cluster energetics at varying levels of He irradiation through functional forms that include contributions from cluster-cluster and cluster-bubble interactions as well as cluster-surface interactions. Such parameterizations of the energetics of helium cluster interactions with tungsten defects and surfaces are subsequently incorporated into hierarchical multiscale models of helium cluster dynamics in PFC materials.

Furthermore, based on detailed characterization of the configurations generated from the large-scale MD simulations, we construct models of plasma-exposed tungsten with a near-surface helium nanobubble region represented by regular arrays of over-pressurized nanobubbles of proper sizes and separation distances. We carry out MD simulations of the deformation state of such model nanobubble regions near tungsten surfaces of various crystallographic orientations. In all cases examined, we identify the state of stress in the near-surface nanobubble layer as equibiaxial compressive stress. The MD-computed biaxial stress level is in good quantitative agreement with that predicted by a heterogeneous elastic inclusion model in conjunction with an equation of state for the pressure in the helium bubbles. Our findings for the state and level of stress in PFC tungsten constitute an important component in the development of atomistically-informed, continuum-scale models of surface morphological evolution in PFC materials.

562791 Thermomechanical Properties of Plasma-Exposed Tungsten

Studying the impact of helium (He) ion implantation on the thermomechanical properties of tungsten is of utmost importance for evaluating tungsten as a plasma-facing component (PFC) in nuclear fusion devices. In a previous study, based on nonequilibrium molecular-dynamics (MD) simulations, we reported that the presence of plasma-related structural defects in tungsten single crystals, namely, nano-scale sized voids and He nanobubbles, causes a significant reduction in tungsten's lattice thermal conductivity, k, down to ~20% of that of perfect crystalline tungsten, with $k$decreasing with increasing void size and increasing He pressure in the nanobubbles.

Here, we report computational results for the thermal expansion coefficient, $\alpha(T,p)$, of defect-free crystalline tungsten over a broad range of temperature, $T(0 \leq T \leq 2000$ K), and pressure, $p(0 \leq p \leq 10$ GPa). Based on isothermal-isobaric (NpT) MD simulations according to a reliable interatomic potential for well-converged supercell sizes, we demonstrate a ~40% reduction in the value of α at 1000 K when the pressure increases from 0 to 10 GPa; hydrostatic stresses of several GPa are expected to develop in plasma-exposed tungsten as a result of He implantation and formation of over-pressurized He nanobubbles. To assess our classical MD predictions, we have carried out first-principles computations of the thermal expansion coefficient of tungsten within density functional theory (DFT). These computations were based on free energy calculations according to density functional perturbation theory (DFPT) for determining vibrational entropies, with the free energies calculated within the quasi-harmonic approximation (QHA). Such first-principles results are used to both validate our classical MD predictions and contribute to the construction of a property database toward classical potential development. These thermal expansion coefficient results also can be utilized as input to continuum-scale models for calculations of thermal stresses in plasma-exposed tungsten and their contribution to the total stress in the plasma-facing material, an important component in developing models of PFC surface morphological evolution.

Furthermore, we report results of systematic MD computations of the elastic modulus of single crystalline tungsten containing nanopores; filling the nanopores with the proper amount of He to form helium nanobubbles is used to generate models of plasma-exposed tungsten and to assess the effect of He-ion irradiation on tungsten elastic properties. We conduct a systematic exploration of parameter space: the parameters varied include porosity (tungsten matrix density), temperature, bubble radius, bubble arrangement in regular arrays, and helium content. Our computations reveal that the dependence of the elastic modulus of nanoporous tungsten on its density follows a power law, similar to that of the modulus-density relation of natural cellular materials. We find that arrangement of voids/bubbles in various lattice configurations in the tungsten matrix does not have an important effect on the elastic modulus. We also find that filling the nanopores with helium increases the stiffness of PFC tungsten and decreases the sensitivity of its stiffness to porosity. At given temperature, we find that decreasing He nanobubble size leads to stiffening of PFC tungsten and reduces the sensitivity of its stiffness to porosity. Finally, we find that at given nanobubble size, raising the temperature leads to relative stiffening of PFC tungsten, with respect to the stiffness of perfect tungsten at each temperature, and reduces the stiffness sensitivity to porosity. Our findings enable the development of a quantitative database for the elastic modulus dependence on He content in PFC tungsten at various plasma exposure conditions, which constitutes another important component toward modeling and simulation of PFC surface morphological evolution.

AUTHOR INDEX

Allred, Jarrod 1
Azzaoui, Taha 46
Battraw, Ben C. 25
Blondel, Sophie 50
Bronikowski, Michael G. 30
Bruffey, Stephanie 5
Burns, Carolyn A. 2, 3
Carney, Andrew 1
Choi, Seungrag 6
Choi, Suhee 20, 22
Clark, Austin 25
Colburn, Heather 1
Cooke, Gary 12
Dale, Olivia 22
Daniel, Richard 2, 3, 4
Dasgupta, Dwaipayan 48, 50, 52
De Almeida, Valmor F. 46
Duan, Yuhua 26, 28
Elsaidi, Sameh 15
Fellinger, Terri 11
Fiskum, Sandra 1
Fitzhugh, Richard L. 21
Fountain, Matthew 2
Grover, Martha A. 8, 14
Haider, Syed I. 23
Hammond, Karl D. 47, 49, 51, 52
Hansen, Avenly H. 25
Harb, John 21
Helal, Ahmed 15
Hopkinson, David 15
Hu, Lin 52
Huang, Zhengliang 16
Jackson, Matt 18
Jia, Ting 26
Jubin, Robert 5
Kalaga, Dinesh V. 24
Kawaji, Masahiro 24
Keefer, Mark 11
Kocevska, Stefani 8, 14
Kolasinski, Robert 48, 50
Ladshaw, Austin 6
Lambert, Dan P. 9
Laurinat, James E. 30
Lee, Wan L. 25
Li, Ju 15
Long, Alexander 18
Lyon, Kevin 46
Maggioni, Giovanni M. 8, 14
Maroudas, Dimitrios 47, 48, 49, 50, 52, 53
McAllister, Justin 17
Memmott, Matthew 21
Meyer, Gabriel 25
Minette, Michael J. 2
Mofrad, Amir M. 51
Mohamed, Mona 15
Monreal, Marisa 18
Nan, Yue 6
Nickerson, Stella D. 25
Page, Jason 12
Parker, Scott 18
Patton, Brad 13
Paudel, Hari 28
Pease, Leonard 2
Peterson, Reid 1, 3, 4
Phillips, Nathan 2
Reynolds, Jacob 12
Robinson, Sharon 13
Rousseau, Ronald W. 8, 14
Rovira, Amy 1
Rudra, Apoorva 24
Schonewill, Philip P. 3, 4
Senor, David J. 26, 28
Serkowski, Jason 2
Shimskey, Rick W. 3
Silker, Aubrey I. 11
Simpson, Michael F. 17, 20, 22
Smoot, Margaret 1
Solano, Andrew R. 25
Song, Qi 16
Spencer, Barry 5
Spicer, Thomas O. 23
Strianese, Andrew 20
Sun, Casper 23
Sun, Jingyuan 16
Tait, Granite J. 25
Tavlarides, Lawrence L. 6
Thallapally, Praveen K. 15
Tsouris, Costas 6
Veldman, Timothy 2
Vogel, Sven 18
Wang, Jingdai 16
Weerasinghe, Asanka 50, 52, 53
Wells, Beric 12
Wiechert, Alexander 6
Williams, Tyler 25
Winkler, Bjorn 18
Wirth, Brian D. 47, 48, 49, 50, 52, 53
Wu, Jianzhong 7
Yang, Yao 16
Yang, Yongrong 16
Yang, Zhuocen 49
Yiacoumi, Sotira 6
Yu, Xiao Ying 2
Zhou, Musen 7